Applications of Modern Mass Spectrometry

(Volume 1)

Edited by

Prof. Atta-ur-Rahman, *FRS*
Kings College
University of Cambridge
Cambridge
UK

Prof. M. Iqbal Choudhary
H.E.J. Research Institute of Chemistry
International Center for Chemical and Biological Sciences
University of Karachi
Karachi-75270
Pakistan

&

Prof. Syed Ghulam Musharraf
H.E.J. Research Institute of Chemistry
International Center for Chemical and Biological Sciences
University of Karachi
Karachi-75270
Pakistan

Applications of Modern Mass Spectrometry

Volume # 1

Editors: Atta-ur-Rahman, M. Iqbal Choudhary & Syed Ghulam Musharraf

ISBN (Paperback): 978-981-14-3380-1

need for a court order if at any point you breach any terms of this License Agreement. In no event will any delay or failure by Bentham Science Publishers in enforcing your compliance with this License Agreement constitute a waiver of any of its rights.

3. You acknowledge that you have read this License Agreement, and agree to be bound by its terms and conditions. To the extent that any other terms and conditions presented on any website of Bentham Science Publishers conflict with, or are inconsistent with, the terms and conditions set out in this License Agreement, you acknowledge that the terms and conditions set out in this License Agreement shall prevail.

Bentham Science Publishers Pte. Ltd.
80 Robinson Road #02-00
Singapore 068898
Singapore
Email: subscriptions@benthamscience.net

BENTHAM SCIENCE

CONTENTS

PREFACE

Mass spectrometry, combined with the modern tools of chemo- and bioinformatics, has emerged as one of the most powerful tools in research with numerous applications in industry. The analytical capacity of mass spectrometry has exceeded beyond its conventional role of determination molecular mass and structural determination. Recent developments in ionization, detection techniques and data analysis tools, as well as applications of mass spectrometry in various fields continue to open up new vistas in this rapidly evolving field. The present volume of *"Applications of Modern Mass Spectrometry"* provides a useful insight into some of these developments. The present 1ˢᵗ volume of this book series comprises of 5 comprehensive reviews, written by the leading practitioners of mass spectrometry. These articles present diverse applications of mass spectrometry in fields such as animal nutrition, food and environmental analysis, biomedical and forensic sciences, as well as toxicology and explosives. Qualitative and quantitative analysis and identification of proteins and peptides are the cross cutting themes in many of these articles.

The review by Gonzalez-Ronquillo *et al.* focuses on the accurate quantification of microbial protein synthesis in ruminants by employing ^{15}N isotopic mass spectrometry. Zhu *et al.* review the recent developments in qualitative and quantitative analysis of proteins and peptides in food matrices through LC-MS techniques. Izadmanesh and Ghasemi have provided an extensive review of the use of chemometric tools for the analysis of complex mass spectrometric data. Heavy metal toxicity is a global health challenge, and reliable methods are required for the detection of heavy metals in water and other samples. Developments in this area employing ICP-MS methods are reviewed by Boruah and Biswas. Finally Anilanmert and Cengiz have reviewed the relevant literature on the direct and on-site detection of various explosives and their residues by modern mass spectrometry. Each article presents case studies of the use of various innovative mass spectrometric tools, along with their strengths and limitations.

We are grateful to all the authors for their excellent scholarly contributions, and for the timely submissions of their review article. We would also like to express our gratitude to Ms. Fariya Zulfiqar (Manager Publications) and Mr. Mahmood Alam (Director Publications) of Bentham Science Publishers for the timely completion of the volume in hand. We sincerely hope that the efforts of authors and production team will help readers in better understanding and appreciating the versatility and robustness of mass spectrometry, and motivate them to conduct good quality research and development work in this exciting area.

Prof. Atta-ur-Rahman, *FRS*
Kings College
University of Cambridge
Cambridge
UK

Prof. M. Iqbal Choudhary
H.E.J. Research Institute of Chemistry
International Center for Chemical and Biological Sciences
University of Karachi
Karachi-75270
Pakistan

Prof. Syed Ghulam Musharraf
H.E.J. Research Institute of Chemistry
International Center for Chemical and Biological Sciences
University of Karachi
Karachi-75270
Pakistan

List of Contributors

Beril Anilanmert Istanbul University-Cerrahpasa, Institute of Forensic Sciences and Legal Medicine, Istanbul, Turkey

Bijoy Sankar Boruah Applied Optics and Photonics Research Laboratory, Tezpur University, Napaam-784028, Assam, India

Ivan Mendez Martinez Departamento de Nutricion Animal, Universidad Autonoma del Estado de Mexico, Instituto Literario 100, 50000, Toluca, Mexico

Jahan B. Ghasemi Faculty of Chemistry, University of Tehran, Tehran, Iran

Joaquim Balcells Departamento de Ciencia Animal, Universidad de Lleida, ETSEA, Alcalde Rovira Roure 191, 25198, Lleida, Spain

Lizbeth E. Robles-Jimenez Departamento de Nutricion Animal, Universidad Autonoma del Estado de Mexico, Instituto Literario 100, 50000, Toluca, Mexico

Lulu Zhao College of Biotechnology of Pharmaceutical Engineering, Nanjing Tech University, Nanjing, 211816, China

Manuel Gonzalez-Ronquillo Departamento de Nutricion Animal, Universidad Autonoma del Estado de Mexico, Instituto Literario 100, 50000, Toluca, Mexico

Rajib Biswas Applied Optics and Photonics Research Laboratory, Tezpur University, Napaam-784028, Assam, India

Salih Cengiz Istanbul University-Cerrahpasa, Institute of Forensic Sciences and Legal Medicine, Istanbul, Turkey

Yahya Izadmanesh Faculty of Chemistry, University of Tehran, Tehran, Iran

Yi-Shen Zhu College of Biotechnology of Pharmaceutical Engineering, Nanjing Tech University, Nanjing, 211816, China

Zhonghong Li Jiangsu Institute for Food and Drug Control, Nanjing, 210019, China

Applications of Mass Spectrometry for the Determination of Microbial Crude Protein Synthesis in Ruminants

Lizbeth E. Robles-Jimenez[1], Ivan Mendez Martinez[1], Joaquim Balcells[2] and Manuel Gonzalez-Ronquillo[1,*]

[1] *Departamento de Nutricion Animal, Universidad Autonoma del Estado de Mexico, Instituto Literario 100, 50000, Toluca, Mexico*

[2] *Departamento de Ciencia Animal, Universidad de Lleida, ETSEA, Alcalde Rovira Roure 191, 25198, Lleida, Spain*

Abstract: The importance of quantifying ruminal microbial crude protein synthesis has promoted the development and comparison of several different methods for precise determination of both the amount and rate of synthesis. One major challenge is in estimating and differentiating protein in the rumen between microbial, dietary, and endogenous fractions, and to correctly isolate the solid and liquid microbial fraction of the rumen contents. This is further complicated by the goal of using non-invasive methods as much as is feasible, such as avoiding the use of fistulated animals; the selection of an appropriate microbial marker, specifically one that behaves similarly in the solid-associated and liquid-associated microbial fractions. It is also vital to be able to accurately estimate the contribution of microbial protein to overall nitrogen used by the animal, which can be accomplished by the use of ^{15}N labeled, as assimilated by ruminal bacteria, and by the quantification of labeled nitrogen *via* mass spectrometry ($^{15}N/^{14}N$). This review focuses on challenges regarding accurate quantification of microbial crude protein synthesis in the rumen, as well as providing the methodology for quantification using the ^{15}N marker. This review is based on the collection of scientific papers from the main research groups in feed and animal nutrition in ruminants.

Keywords: Endogenous excretion, ^{15}N, Microbial protein, Purine derivatives, Ruminants.

INTRODUCTION

Ruminants are inefficient utilizing dietary nitrogen, so they have to use microbial protein (MP) to meet their metabolizable protein requirements [1, 2]. In order to

* **Corresponding author Manuel Gonzalez-Ronquillo:** Departamento de Nutricion Animal, Universidad Autonoma del Estado de Mexico, Instituto Literario 100, 50000, Toluca, Mexico; E-mail: mrg@uaemex.mx

Atta-ur-Rahman, M. Iqbal Choudhary & Syed Ghulam Musharraf (Eds.)

know the crude protein (CP) requirements of ruminants and improve their efficiency, microbial crude protein (MCP) synthesis is used. Menezes *et al.* [3] mentions that if more CP is given than required by the ruminant, the excreted Nitrogen (N) increases instead of improving their performance.

The importance of quantifying MCP in the rumen has spurred the development and comparison of several different methods of analysis. However, these methods do not always satisfy the required scientific needs for specificity, efficiency, or cost, nor do they always address concerns regarding quantitative inconsistency in replicating this technique for estimation. Some of the major challenges of these methods focus includes 1) how to estimate and differentiate protein of microbial, dietary, and endogenous origins; 2) proper isolation of the microbial fraction in the particulate and fluid fractions of the rumen; 3) how to accurately quantify microbial protein in a minimally invasive manner, such as without the use of fistulated animals (ruminally and duodenally); and 4) the choice of an appropriate microbial marker that behaves similarly in both solid and liquid fractions [4, 5]. Regarding this final challenge, a currently used microbial marker is the labeled nitrogen ^{15}N, which is one of the most reliable and recommended methods, unlike others than often overestimate or underestimate MCP [6 - 10]. This review focuses on discussion of these challenges regarding accurate quantification of MCP synthesis in the rumen, as well as providing the methodology for quantification using the ^{15}N marker. This review based on the collection of scientific papers from the main research groups in feed and animal nutrition in ruminants.

MICROBIAL CRUDE PROTEIN SYNTHESIS

Nutritional studies in ruminants are aimed at the selection of feeds based on high efficiency of MCP synthesis in the rumen along with the available N sources and energy support. A key strategy for improving production has, therefore, been designed to maximize the efficiency utilization of available feed resources in the rumen by providing optimum conditions for microbial growth and thereby, supplementing dietary nutrients to complement and balance the products of rumen digestion to the animal's requirement. Supplementation with rumen-protected lysine and methionine can improve N use efficiency, maximizing the CP requirement from 18 to 15% in dairy cattle, without affecting milk yield production or animal performance [11].

Feed consumed by the ruminant, such as forage and cereals, enters the rumen and is available for degradation by rumen microbes. One fraction is rapidly degraded (Fraction A), while another is degraded slowly (Fraction B). Both fractions are then used for microbial growth and synthesis of MCP (Fig. **1a**) [12]). This process depends on several factors, such as type of feed, ratio of forage: concentrate in the diet, fractional rate of degradation in the rumen, physiological stage of the animal, presence of secondary compounds (saponins, tannins, polyphenols, *etc.*), as well as the use of additives, such as enzymes or ionophores, all of which can affect digestion and microbial kinetics.

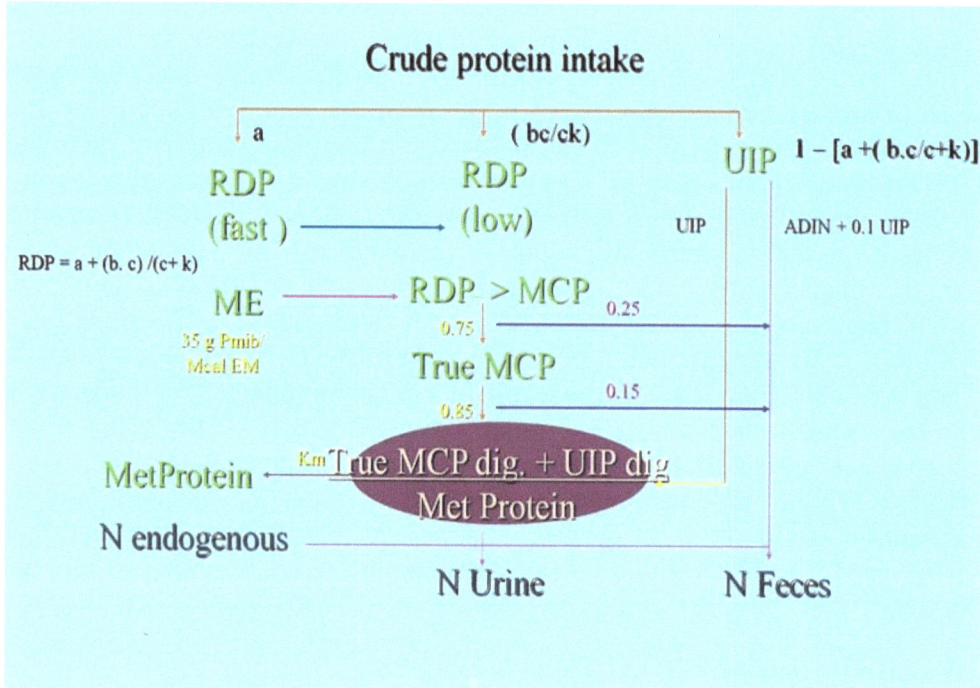

Fig. (1a). Ruminal degradable fractions, and microbial crude protein synthesis into the rumen (adapted from AFRC [12]).

In addition to the factors mentioned above, the synthesis of MCP is dependent on the energy supplied, which averages 35 g MCP/Mcal of metabolizable energy (ME) intake across particulate-associated bacteria (PAB) and liquid-associated bacteria (LAB). If the diet is energy deficient, there will be a corresponding reduction in MCP synthesis and ammonia from breakdown of amino acids and NPN will be absorbed into the bloodstream rather than being used for formation of MCP [13]. If the diet is too high in protein, excess protein will also be converted to ammonia and absorbed across the rumen wall into the bloodstream [14].

Once this MCP has been synthesized in the rumen, depending on the method of estimation [15], 75% is true microbial protein, the rest (25%) are nucleic acids, 85% is true digestible microbial protein (15% are undegradable amino acids) therefore, for every 100 g of MP synthesized in the rumen, only 64 g will reach duodenum in the form of truly digestible MCP (Fig. **1b**). Dietary protein, undegradable in the rumen (RUP), will exit to the duodenum. The amount of true RUP varies depending on diet and processing methods, though the calculation of RUP includes Acid detergent insoluble nitrogen (ADIN). Protein reaching the duodenum is of both sources, food and microbial origin, and is called protein digestible in the intestine (PDI). However, the amount of MCP varies depending on the factors discussed above, hence it is important to quantify the protein flow entering the duodenum, and to differentiate into that of microbial or food origin, combined with the Non-Ammonia Nitrogen (NAN) that reaches the duodenum, and is recycled by the ruminant [15 - 17].

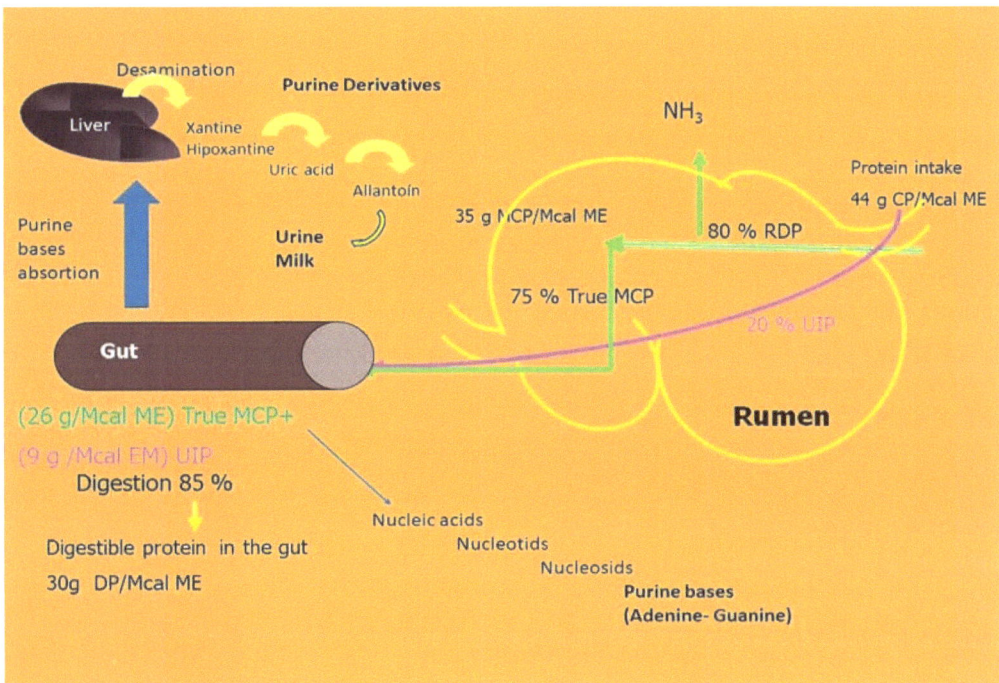

Fig. (1b). Microbial crude protein synthesis into the rumen and the estimation of microbial protein through purine derivatives excretion in dairy cattle.

DIETARY NITROGEN FRACTIONS AND THEIR POTENTIAL USE BY RUMINANTS

The Cornell Net Carbohydrate and Protein System (CNCPS) is made up of a series of sub-models that assess, respectively, the carbohydrate and protein content available in the diet [18], processes of fermentation and MP synthesis [19], energy and protein requirements of cattle [20], and needs of essential amino acids [21] (*i.e.* lysine and methionine). The CNCPS predicts the requirements, feed utilization, animal performance, and nutrient excretion for dairy and beef cattle using accumulated knowledge about feed composition, digestion, and metabolism in supplying nutrients in order to meet the nutritional requirements.

Dietary protein is traditionally divided up into true protein (amino acids) and non-protein nitrogen (NPN), but Licitra *et al.* [22] developed a new concept to differentiate nitrogen fractions present in feed and their potential uses by ruminants (Table **1**). In the CNCPS, three nitrogen fractions are differentiated: non-protein N (NPN, fraction A), which is used exclusively in the form of $N-NH_3$, the potentially degradable true protein (fraction B) and the protein bound to the detergent acid fiber (fraction C), which is not digestible in the intestines. Moreover, fraction B is subdivided into three others that are characterized by their different degradation rates, as indicated in Table **1**. Considering the degradation (k_d) and passage (k_p) rates allows the estimation of the N contribution available to microorganisms in form of $N-NH_3$ and peptides or amino acids, as well as the proportion of protein that escapes without being degraded (UIP).

As mentioned earlier, there is a first fraction that quantifies the non-protein nitrogen (NPN, Fraction A) and the true protein; true protein is divided into a rapidly soluble (BSP, Fraction B_1) and another that is insoluble but potentially degradable (smaller particles) which is called fraction B, the N fractions that are bound to dietary fiber (hemicellulose, cellulose, and lignin), Fraction B_2 is linked to Neutral Detergent Fiber, fraction B_3 is soluble in Acid Detergent Fiber and finally, there is a fraction insoluble in acid detergent (Fraction C), which is known as ADIN. The rapidly soluble fractions (Fraction A), together with the potentially degradable fractions in rumen (Fraction B), can be degraded by rumen bacteria and synthetize MCP.

Table 1. Partition of nitrogen and protein fraction in feedstuffs.

Fraction	Abbr.	Estimation or Definition	Enzymatic Degradation	Classification[a]
Nonprotein N True protein	NPN TP	Not precipitable Precipitate with tungstic acid	Not applicable	A
True soluble protein	BSP	Buffer soluble but precipitable (TP-IP)	Fast	B_1
Insoluble protein Neutral detergent soluble protein	IP IP-NDIP	Insoluble in buffer Difference between IP and protein insoluble in neutral detergent (ND)	Variable	B_2
ND insoluble protein, but soluble in AD	NDIP-ADIP	Protein insoluble in ND but soluble in acid detergent (AD)	Variable to Slow	B_3

(Table 1) cont.....

Fraction	Abbr.	Estimation or Definition	Enzymatic Degradation	Classification[a]
Insoluble in acid detergent	ADIP or ADIN	Includes heat-damaged protein and nitrogen associated with lignin	Indigestible	C

[a] Adapted from Licitra *et al.*, [22].

METHODS OF DIRECTLY AND INDIRECTLY ESTIMATING MICROBIAL PROTEIN SYNTHESIS

Interest in estimation of MCP synthesis goes back to the study of Faichney in 1975 [14] who estimated the duodenal flow of purine bases (*i.e*, adenine and guanine), from the content of nucleic acids present in bacteria (DNA, RNA), which contain purine bases (PB). Other authors used some markers, such as diaminopimelic acid (DAPA) and D-Alanine, which are contained in the bacterial fraction of the rumen, while other authors calculated the PB / N ratio, based on PAB and LAB, with variable results [23, 24]. Therefore, in recent years, the isotopic enrichment of ruminal bacteria and purine bases with $(^{15}NH_4)_2SO_4$ and $^{35}SO^4{}_{-2}$ was chosen for more general use, due to its improved accuracy [25, 5].

Likewise, indirect non-invasive methods have been developed that allow the estimation of MCP synthesis from the excretion of purine derivatives (PD, *i.e*. Xanthine, Hypoxanthine, Uric acid, and Allantoin) in urine and milk [26], with favorable results found in urine (Fig. **2**).

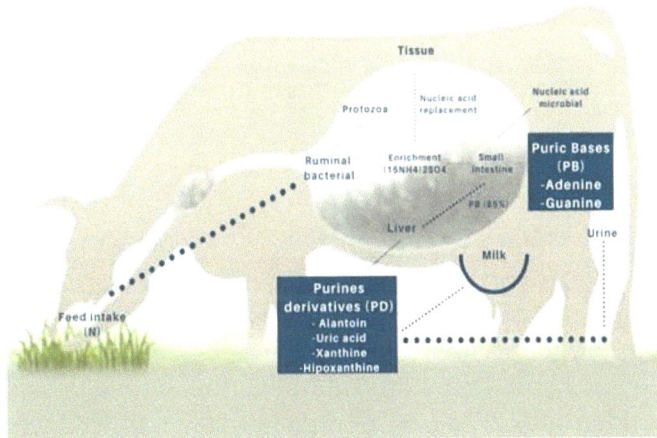

Fig. (2). Estimation of microbial protein through purine derivatives excretion in dairy cattle.

CHARACTERISTICS AND USES OF ^{15}N

Nitrogen labeled as ^{15}N is an example of an *external marker*, which is administered to animals in an inorganic form. These inert markers are then incorporated into the microbial mass in the rumen, allowing for the differentiation the dietary, endogenous, and microbial fractions of rumen contents [26]. According to Carro [27], and Broderick and Merchen [23], these markers are easy to administer to the animal, and provide a constant enrichment rate, without risks of contamination or radioactivity – they specifically enrich ruminal microbes, rather than marking feed particles, and allow for the differentiation of microbial versus dietary protein. Such labeled nitrogen is directly incorporated into the rumen's bacterial populations, and indirectly into rumen protozoa during protozoal predation of other rumen microorganisms [23]. These specific characteristics allow for the use of external markers, such as ^{15}N, in *in vivo* or *in vitro* studies, with a high reliability rate.

Common Uses of ^{15}N

The labeled nitrogen ^{15}N has been used in studies focusing on microbial protein synthesis as an indicator of efficiency on diets that promote optimal maintenance and health of rumen microorganisms [10, 28 - 30]. This aspect is particularly important, as microbial protein, a major source of protein available to the ruminant, accounting for 60 to 90% of protein entering the small intestine, in contrast to dietary protein [28, 1], though this proportion is not constant and varies depending on a combination of factors, including dietary and animal factors.

Another alternative for the administration of ^{15}N to ruminants is from the production of ^{15}N labelled feed during growth (forages). For this purpose, $^{15}NH_4^{15}NO_3$ (Larodan Fine Chemicals AB, Malmö, Sweden) containing 2% of $^{15}N/^{14}N$, $(NH_4)_2SO_4$ with 10% enrichment of $^{15}N/^{14}N$ (Isotec, Miamisburg, OH) is used. The animals then are fed with this forage, both silage and hay enriched with ^{15}N [31 - 33].

^{15}N-labelled Forage

For the cultivation of forages (*i.e.* Alfalfa, Timothy grass) labeled with ^{15}N are fertilized with 1,100 g of $(NH_4)_2SO_4$ label. The grass is then cut, withered for 3 hours at 20°C and subsequently ensiled, in addition to using formic acid (5 mL/kg) as a preservative. The silage is defrosted at room temperature and separated into fractions enriched with soluble and insoluble substances. Portions of 20 g are suspended in 400 mL of ultrapure water (Milli-Q, Merck Millipore Corporation, Darmstadt, Germany) and stirred for 1 h at 39°C. Then, the suspension is centrifuged at 15000 x g for 15 min (Avanti J26S XP, Beckman Coulter, Inc. Brea, CA, USA) and filtered through Whatman no. 1 filter paper. The filtrate consisting of the soluble fraction, subsequently frozen at -20 °C and freeze-dried (Fig. **3**) ([32]) [32, 33].

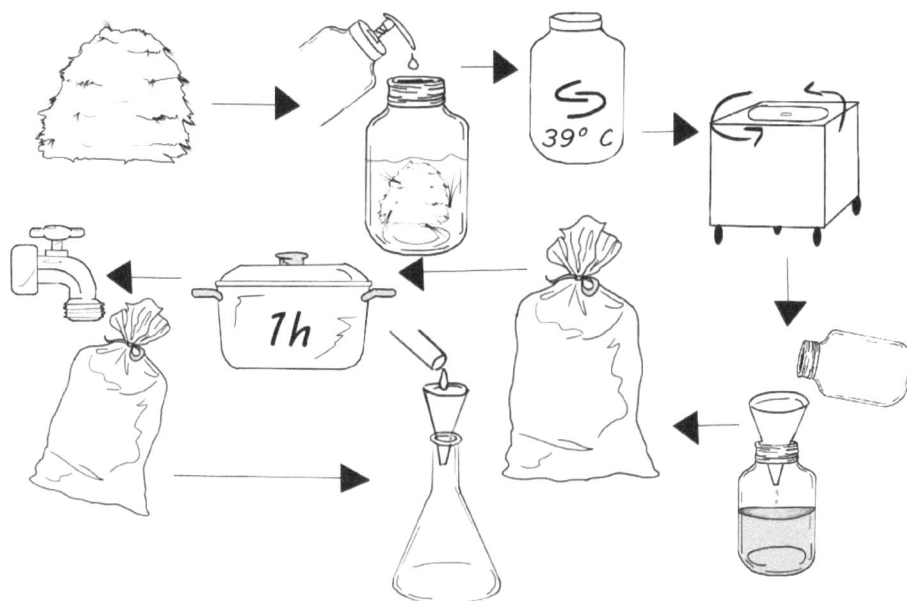

Fig. (3). Procedures used for extracting forage N fractions from silages and dried grass (adapted from Vaga *et al.* [32]).

Two samples of the soluble fraction are taken for chemical analysis. The first 15 mL sample is treated with 0.3 mL of 50% (vol/vol) H_2SO_4 to determine the concentration of ammonia N (AN) marked with ^{15}N and the excess ammonia N (APE) at 15%. A further 200 g of soluble fraction is treated with 5 mL of saturated $HgCl_2$ solution for the analysis of total soluble N, and the APE of total soluble N. To calculate the APE in each fraction, the ^{15}N-atom% background is determined from unlabeled samples. All samples are stored at -20°C until further analysis.

Forage labelled ^{15}N is administered to ruminants on days 10 and 11 of the experiment. For ruminal metabolism of ^{15}N fractions, samples are taken from ruminal digestion at different times, 0.25, 0.5, 0.75, 1.0, 1.5, 2, 3, 4, 6, 8, 11, 14, 17, 22, 27, 33, 39, 47, 55, 63, and 72 h after administration of N-labeled sources. For more information, see Ahvenjärvi *et al.* [33].

Ammonium Sulfate & Ammonium Chloride

Ammonium sulfate($^{15}NH_4SO_4$) and ammonium chloride ($^{15}NH_4Cl$) are the most commonly used forms of ^{15}N used for direct applications [23], with ammonium sulfate being used more frequently [10, 29, 31, 34, 35]. The typical procedure for use of ^{15}N involves the administration of the labelled compound *via* continuous ruminal infusion, allowing for uniform enrichment of rumen microbes. The advantage offered by these compounds is that they dissolve rapidly in water, allowing for the integration of the ^{15}N over a short time, approximately 3 days [29, 31, 35].

Omasal Sampling and Analysis

The omasal sampling technique can be an alternative for the calculation of the MCP synthesis, being a less invasive method and allowing to investigate in more detail the flow of LAB, PAB and NAN components in the rumen, if duodenal cannulated animals are not available [36].

This procedure is done during the last week of each experimental period (when sampling omasal samples for MCP synthesis) using the techniques developed by Huhtanen *et al*. [37] and Ahvenjärvi *et al*. [38], as adapted by Reynal and Broderick [1], to quantify digesta flow from the rumen of the indigestible NDF [39], $YbCl_3$ [40], and Co-EDTA [41]. Approximately 500 mL of omasal digesta is sampled at each sampling point (n= 3). A sub-sample (70 mL) is used to obtain the LAB and the PAB. Another sub-sample (approximately 400 mL) is frozen −20 °C. This sample is filtered in cheesecloth and washed with 1 L of 0.9% NaCl solution. The solids retained in the cheesecloth are mainly associated with the large particle (LP). The filtrate is centrifuged (1000Å~g for 10 min at 5 °C) and the precipitate is considered small particle (SP), whereas supernatant is designated fluid phase (FP) of digesta [37], respectively, are used as flow markers in omasal or duodenal contents. Indigestible NDF is determined in LP, SP, and total mixed ration (TMR) but not in FP [38]. During the determination of indigestible NDF, samples (0.35g) are weighed into duplicate 5 x 10-cm dacron bags with 6-μm pore size, incubated in the rumen of two or three dairy cows for 12 days, rinsed with water, then subjected to NDF analysis, as described below. The external microbial marker ^{15}N is used to quantify NAN flow from the rumen.

The triple marker technique [42, 43] used to determine the proportions with which to recombine the 3 phases of rumen contents to quantify omasal true digesta (OTD). Before marker infusion begins, whole ruminal contents are taken from each cow to determine the background ^{15}N abundance (Natural abundance). Means of the total observations of background ^{15}N abundance vary (*i.e.* 0.3681% of N). Cobalt-EDTA, $YbCl_3$, and $^{15}NH_4SO_4$ containing 10% atom excess ^{15}N (*i.e.* Isotec, Miamisburg, OH) are dissolved in distilled water and continuously infused into the rumen at rates of 2.0 g of Co EDTA, 3.0 g of Yb, and 70 mg of ^{15}N per day for approximately 6 days. Markers are continuously infused for 6 days using a peristaltic pump. After 3 days of infusion, ruminal and omasal samples are collected at 2-h intervals over a 3-d period to represent the 24-h flow. The sampling protocol includes confirming that sample tubes are correctly positioned in the omasal canal, sampling times and volumes, sample processing, isolation of LAB and PAB, digesta marker analyses, and preparation of omasal true digesta, as described by Reynal and Broderick [1] and Brito *et al.* [44], except that ammonia and protozoa are not isolated for determination of ^{15}N enrichment. Samples of OTD are analyzed for total N, DM, OM, NDF, ADF, NPN, and ADIN. Samples of OTD and isolated bacteria are treated with K_2CO_3 [44] to remove residual ammonia and analyzed for total N (equivalent to NAN) and for ^{15}N abundance using an elemental analyzer (*i.e.* Costech Analytical Technologies Inc., Valencia, CA) interfaced to a Thermo-Finnigan Delta-Plus Advantage isotope ratio mass spectrometer (*i.e.* Thermo-Electron GmbH, Bremen, Germany). Equations used to compute nutrient flow of dietary and microbial origin and extents of ruminal digestion are described by Brito *et al.* [44].

Feed samples, ruminal content, and OTD from any experiment, in order to determine MCP synthesis in ruminants, are determined by drying at 60 °C (forced-air oven) for 48 h. Feed samples, ruminal content and OTD are ground to pass a 1-mm Wiley mill screen, and analyzed later for total N (*i.e.* Leco FP-2000 Nitrogen Analyzer, Leco Instruments Inc., St. Joseph, MI), DM, OM [45], and then sequentially for NDF, ADF, and ADIN using heat-stable α-amylase and Na_2SO_3 [46] as per Hintz *et al.* [47], as well as for NPN without use of Na_2SO_3 [22].

0.5 g sample of OM truly digestible in the rumen (OMTDR) from each cow or any ruminant content, which is extracted in 10 mL of citrate buffer (77.5 mM adjusted to pH 2.2 with HCl) for 30 min at 39 °C, and centrifuged at 15,000 x g at 4 °C for 15 min. Broderick *et al.* [48] conducted a meta-analysis of the omasal sampling technique finding that models based on OM intake are better predictors of microbial N flow.

Ruminal and Duodenal Samples

To accurately estimate microbial protein synthesis, it is necessary to calculate the flow of total nitrogen, food nitrogen, microbial N flow (non-ammoniacal non-microbial, N- NANM) in the duodenum [15, 49, 50]. This typically involves the use of ruminally and duodenally cannulated animals, from which samples of the microbial fractions are taken, as well as from of the omasal canal (Rumen), if duodenal cannulas are not available [1] (See above), or in the small intestine at the duodenal level [26] with either T-shaped duodenal cannula, and the inclusion of flow markers (Cobalt-EDTA, Cr-EDTA, $YbCl_3$, Titanium oxide, NDF indigestible) [26, 51].

After samples of digesta are obtained [52], the next step is to differentiate nitrogen by its origin into microbial, dietary, and endogenous pools. For this, the marker / nitrogen ratio in duodenal flow, as well as in ruminal microorganisms present in the PAB and LAB, are calculated.

When performing this procedure, there are differences in the relationships between bacteria associated with PAB and LAB of the ruminal content [1, 25, 34, 53, 54]. The relationship between both values (marker-nitrogen / duodenal flow and marker-nitrogen / bacteria (LAB and PAB)), represents the nitrogen fraction of the flow that is of microbial origin. The labeled ^{15}N is commonly used for these procedures, as it is highly recommended for its accuracy [55, 56].

The *in vivo* techniques for calculating microbial N flow, as described above, have major disadvantages and are also controversial for animal welfare reasons, since cannulated animals are needed, which generate labor costs at the time of surgery and maintenance of the animals [7].

Liquid-associated and Solid-associated Microbes in the Rumen

As previously discussed, the bacteria present in the rumen are associated with the liquid (LAB) and particulate (PAB) fractions, these fractions must be separated and quantified (Fig. **4**). This procedure is performed *via* filtration and rinsing with solutions based on sodium chloride (NaCl) and at low temperatures (2-3 °C) [34]. To separate the supernatant of ruminal fluid and rinse the solution from the bacterial component, various centrifugation techniques are used to reduce bacterial cell rupture [10, 35, 34]. Finally, the bacterial sediment (pellet) passes through a lyophilizate to be analyzed [10, 34]. Mass spectrometry is one of the most commonly used methods to analyze the enrichment of ^{15}N within the bacterial population [10, 57].

Endogenous Excretion of Purine Derivatives

Once the N is absorbed in the small intestine, and is fixed in the tissues, or is synthesized to produce milk, a part of this N that it is not used, is excreted in the urine. Absorbed nitrogen that is not retained nor excreted in milk is excreted as waste. However, waste also contains nitrogen from endogenous sources, which is difficult to quantify.

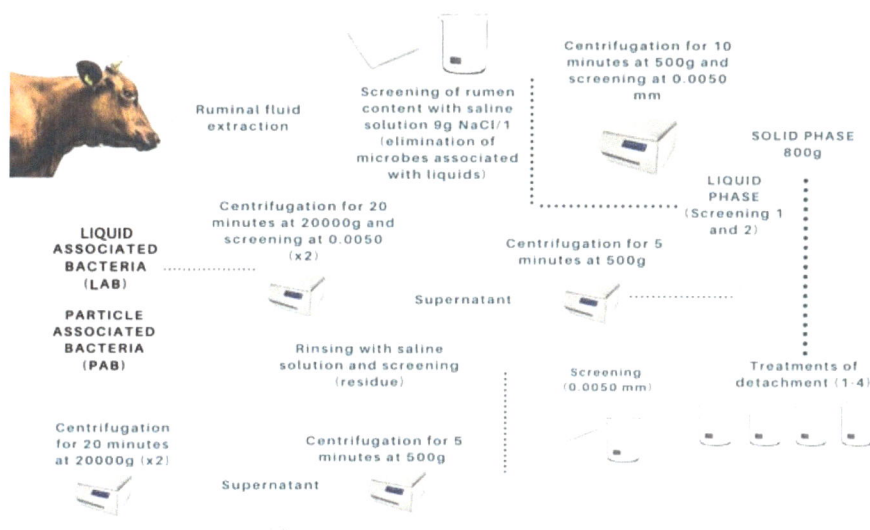

Fig. (4). Isolation of liquid (LAB) and particulate associated bacteria´s (PAB) fractions in ruminal content.

Various techniques have been proposed to help quantify endogenous nitrogen excretions, among them the proposed fasting of the animals [58, 59], or making these animals artificially "non-ruminant", by feeding them on milk. However, the latter method presents difficulty in estimation, because part of this milk ferments in the rumen. Other methods used for quantification of endogenous nitrogen are based on intragastric infusion of nucleic acids [26] or the replacement of duodenal content [60], in animals fistulated in the duodenum with re-entry cannulas, which ultimately affect the normal physiology of the animal. Endogenous nitrogen excretion values have been determined in intra-gastrically fed animals [61, 62], in animals whose duodenal digesta was substituted with an infusion solution containing free purine bases [63], as described Gonzalez Ronquillo *et al.* [26] or following rumen emptying [64]. It is intrinsically assumed in existing reports that endogenous PD losses are equivalent to the basal excretion when there is no exogenous PB absorption from the small intestine. Therefore, it is possible to opt for the continuous infusion of ^{15}N-NH_3 provided naturally, labeled PB in the digesta leaves the rumen in cattle and with it the quantification of the endogenous N fraction of the animals.

In theory, microbial PB absorbed from the small intestine is metabolized and excreted in urine proportionate the rate of flow into the small intestine. However, the relationship between duodenal input and urinary output of purine compounds is masked by the presence of an endogenous fraction originating from the animals *via* their own nucleic acid turnover. Therefore, methods based on the flow of PB to the intestine have been used to quantify the urinary endogenous excretion of PD.

The endogenous urinary PD fraction is determined from the isotopic enrichment of urinary PD and that of duodenal PB (Fig. **5**). Duodenal PB labeled by continuous infusion of ^{15}N supplied as ammonium phosphate. The duodenal flow of PB is determined using a double [65] or triple marker system [39], using Ytterbium acetate and Cr-EDTA, or any other particulate and liquid markers, respectively.

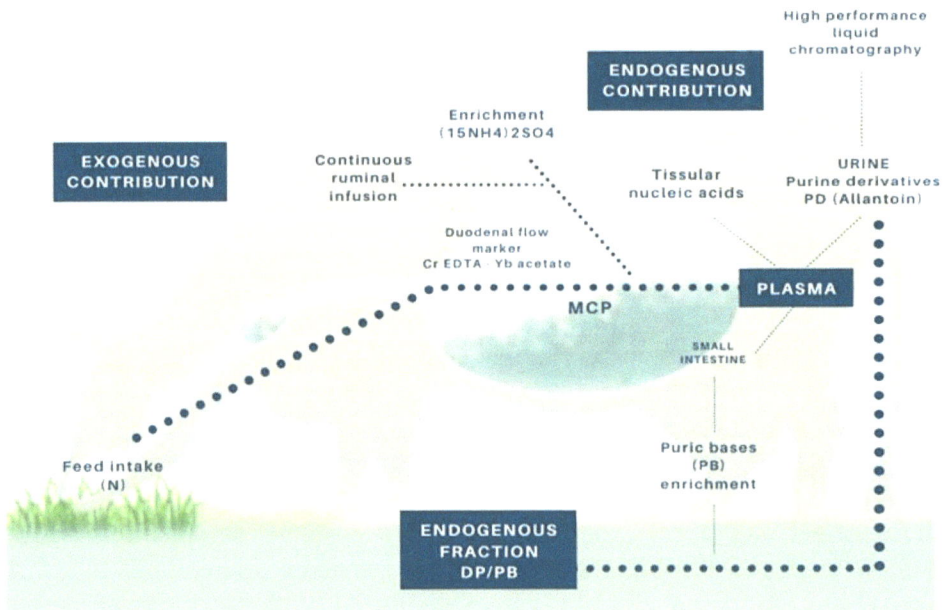

Fig. (5). The endogenous urinary purine derivatives fraction is determined from the isotopic enrichment (^{15}N) of duodenal purine bases enrichment and urinary purine derivatives excretion in urine.

Once $(^{15}NH_4)_2SO_4$ has been infused for five days at least, urine samples and rumen and /or omasal contents are taken from day 6 to 8, and samples of duodenal digesta (250 mL) are collected at 6 hourly intervals for 48 h, refrigerated and pooled on an individual basis at the end of the collection period. Samples of whole and solid digesta are then freeze-dried for subsequent analyses [26]. Urine is collected daily into 200 mL of sulphuric acid solution (10% v/v) by means of external separators glued to the vulva of the dairy cattle or using urethral catheters. Urine is weighed and specific gravity is recorded or weighted, pooled, and stored at –20 °C in order to proceed with the determination of Purine derivatives (PD) and N content. PD in acidified urine is determined by HPLC [66]. Adenine and guanine in duodenal digesta samples (40 mg) are determined by HPLC technique, after acid hydrolysis [67].

Extraction of Purine Bases and Derivatives

Purine bases in duodenal digesta are extracted by specific precipitation with silver ion as an Ag complex following the procedure of Kerr and Seraidarian [67] and modified by Aharoni and Tagari [30]. In this modification, acetic acid is substituted for ammonium-phosphate buffer in order to avoid $^{14}N/^{15}N$ contamination. The purity of the precipitate must be tested by HPLC [66] by screening the effluent from 205 to 300 nm wavelength.

Extraction of Allantoin or Allantoic Acid

Urinary allantoin is extracted by ion-exchange chromatography after its conversion to allantoic acid by alkali hydrolysis. The extraction procedure is as follows. Urine (2 mL) is mixed with 0.5 mL of 0.5 M NaOH and heated on a boiling water bath for 20 min. After cooling in running tap water, the pH of the solution is adjusted to 9.0 and injected into the ion-exchange column (12 x 2 cm; Amberlite IRA-400, Sigma Co., Ma, USA) that has been conditioned previously for 2 h with boric acid buffer (0.05 M boric acid adjusted to pH 9.0 with NaOH).The column then is washed with the same buffer for 2 h and adjusted to pH 6.0 with $(NH_4)H_2PO_4$ (0.05 M, pH 6.0). Ammonium phosphate buffer is allowed to flow for 45 min, approximately (depending the fluid rate), and allantoic acid is eluted using NaCl (1M) in 2 mL-fractions. Chromatography is carried out at room temperature and at a flow rate of 0.6 ml/min (recommended). Allantoic acid is eluted after 25 min of saline elution and its presence is colorimetrically detected using 0.5 ml sub-samples [68]. Allantoic acid fractions are mixed (6 or 8 mL), freeze-dried, and stored until^{15}N analysis. Purity of the eluted samples is tested by re-diluting samples of the freeze-dried material, injecting the resultant solution on the HPLC system [66] and screening for the presence of interfering compounds from 205 to 320 nm wavelength.

Isotope Abundance

The ^{15}N abundance in samples (Urine, LAB, PAB, Duodenal content or omasal content) is determined using a gas isotope-ratio mass spectrophotometer (VG PRISM II, IRMS) in series with a Dumas-style combustion N analyzer (EA 1108 CARLO ERBA). Enrichment of urinary and duodenal PB, during the period of ^{15}N infusion, is corrected for background enrichment in similar samples collected before the infusion was started.

The endogenous fraction is determined by difference in the isotopic enrichment between duodenal and urinary PD. Duodenal PB labelled by continuous infusion of the isotope (^{15}N) supplied as ammonium sulphate (Fig. **6**). The time necessary for equilibrium of labelled purine compounds through the digestive and metabolic pathways should be determined previously, assuming that 48-72 h would be needed to obtain a homogeneous distribution of the isotope within the rumen ecosystem and that a further 24 h would be required for urinary PD enrichments to equilibrate with duodenally absorbed PB [34]. The percentage of urinary PD coming from tissue nucleic acids at steady-state conditions is calculated as follows:

Endogenous excretion of PD (%) = (1 − [^{15}N urinary PD enrichment (allantoin or allantoin precursors) /^{15}N duodenal PB enrichment]) × 100   (1)

% Microbial N = (^{15}N duodenal PB enrichment - natural ^{15}N duodenal PB)/(^{15}N microbial enrichment - ^{15}N microbial natural)) x 100   (2)

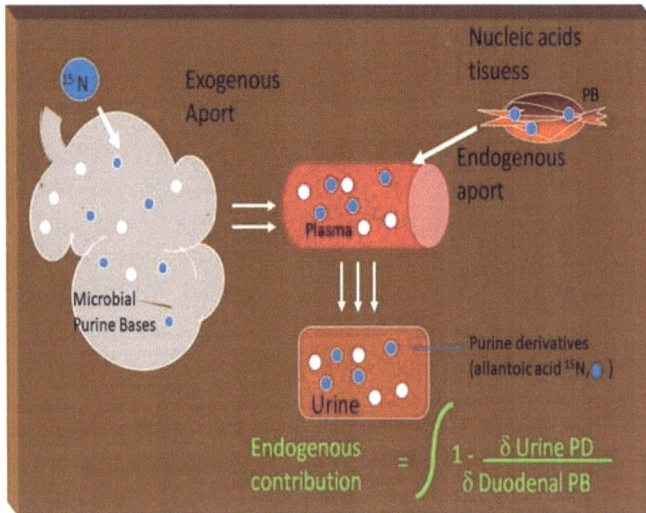

Fig. (6). Endogenous urinary purine derivatives fraction, determined from the isotopic enrichment (^{15}N) of urinary PD (allantoic acid-^{15}N) in urine.

For calculate the Rumen Pool Sizes, Rumen soluble N (RSN) pool size is calculated based on the concentration of soluble N in rumen digesta grab samples and rumen fresh matter pool size determined as a mean of three rumen equations:

Rumen soluble N pool size (g) = rumen fresh matter pool size (kg) × soluble N concentration in rumen digesta (g/kg).   **(3)**

The rumen pool size of ammonia N (AN) and fraction B_1 is calculated similarly to Equation 3. Rumen Soluble Nitrogen Ammonia nitrogen (SNAN) pool size is calculated as a difference between rumen soluble N pool size and rumen AN pool size.

Rumen ^{15}N Pool Sizes in Excess of Background Levels

The background ^{15}N-atom% observed in rumen microbial and digesta samples, taken before dosing ^{15}N, are used to calculate APE in different N fractions. The pool size of rumen soluble ^{15}N in excess of background levels (^{15}NEP) is calculated as:

Rumen ^{15}NEP of soluble N (mg)= rumen soluble N pool size (mg) × ^{15}N excess atom % in rumen soluble N/100.   **(4)**

Rumen ^{15}NEP of AN and fraction B_1 is calculated similarly to Equation 4, and ^{15}NEP of rumen SNAN is calculated by difference between ^{15}NEP of soluble N and AN.

Rumen insoluble N pool size (bacteria N + solids N) is calculated based on rumen insoluble DM pool size determined as a mean of rumen (equation 3) and insoluble N concentrations analyzed from rumen samples.

The APE of rumen insoluble N is not analyzed directly, but instead is calculated as a weighted mean of APE of bacteria N and solids N isolated from rumen digesta grab samples. The fractional proportions of unlabeled feed N, bacterial N, and protozoal N in rumen insoluble N pool size are predicted based on assumption that APE of rumen insoluble N is a weighted mean of each sub-fraction contributing to APE.

Rumen insoluble N (APE) = a × feed N (APE) + b× bacterial N (APE) + c × protozoal N (APE),   **(5)**

Where, a, b, and c are fractional proportions of feed N, bacterial N, and protozoal N in rumen insoluble N pool size. The sum of a + b + c is constrained to 1. Because all APE given to the cows is soluble, APE of insoluble N of feed origin should be 0 at all sampling times, and consequently, all APE of the insoluble N should, therefore, originate from bacterial and protozoal pools. The regression coefficients a, b, and c are iteratively adjusted using the Solver tool in MS Excel (Microsoft Corp., Redmond, WA) until the minimum sum of squares for the difference between observed and predicted APE of rumen insoluble-N (SS_{diff}) is reached.

$$SS_{diff} = \Sigma \text{ (observed APE in rumen insoluble N at time t } - \text{ predicted APE in rumen in soluble N at time t)}^2 \quad (6)$$

Rumen ^{15}NEP of bacterial and protozoal N is calculated based on the regression coefficients, rumen insoluble N pool size, and APE in bacterial and protozoal N:

$$\text{Rumen } ^{15}\text{NEP of bacterial N (mg)} = \text{proportion of rumen bacteria in rumen insoluble N} \times \text{rumen insoluble N pool size (mg)} \times \text{APE in bacterial N/100,} \quad (7)$$

$$\text{Rumen } ^{15}\text{NEP of protozoal N (mg)} = \text{proportion of rumen protozoa in rumen insoluble N} \times \text{rumen insoluble N pool size (mg)} \times \text{APE in protozoal N/100 [33]} \quad (8)$$

Also, the ^{15}N enrichment and the PB concentration in the samples can used to estimate the microbial synthesis. The flow of NAN associated with liquid associated bacteria (LAB-NAN, g/d) and particles adhered bacteria (PAB-NAN, g/d) is calculated as:

$$\text{LAB - NAN} = \text{FP}_{flow} \times \text{NAN FP} \times (\text{FP}_{marker}/ \text{LAB}_{marker}) \quad (9)$$

$$\text{PAB - NAN} = \text{SP flow} \times \text{NAN SP} \times (\text{SP}_{marker}/ \text{PAB}_{marker}) + \text{LP}_{flow} \times \text{NAN}_{LP} \times (\text{LP}_{marker} / \text{PAB}_{marke}) \quad (10)$$

Where, FP_{flow}, SP_{flow}, and LP_{flow} are the flows of omasum phases (g/d); NAN_{FP}, NAN_{SP}, and NAN_{LP} are the non-ammonia nitrogen of omasum phases (g/g); FP_{marker}, SP_{marker}, LP_{marker}, LAP_{marker}, and PAB_{marker} are the microbial markers of omasum phases and bacterial pellets.

Estimates of the endogenous contribution to PD excretion were obtained as following [26, 29]:

$$\text{Endogenous contribution} = \{ 1 - \text{Urine PD } ^{15}\text{N enrichment/ Omasum PB } ^{15}\text{N enrichment}\} \quad (11)$$

CONCLUSION

The use of ^{15}N as a microbial marker has allowed a more precise quantification of microbial protein synthesis in particulate and fluid associated fractions, and the amount of protein of microbial origin that reaches the duodenum versus that of food origin and non-ammonia nitrogen. Likewise, the estimation of endogenous N excretion from the excretion of purine derivatives in urine allows for exploration of the relation of endogenous N from bacteria and the total N excreted in a non-invasive manner. However, techniques for direct determination of the microbial synthesis in the rumen are laborious, expensive, and even difficult to understand, more studies are needed that address the entire procedure for calculating the MCP in ruminants in a didactic way.

CONSENT FOR PUBLICATION

Not applicable.

CONFLICT OF INTEREST

The authors confirm that this chapter contents have no conflict of interest.

ACKNOWLEDGEMENTS

Miss. Lizbeth Robles was benefited by a grant of the Conacyt during her studies of Doctorate, in the program of "Doctorado en Ciencias Agropecuarias y Recursos Naturales", Universidad Autonoma del Estado de Mexico. This study was supported by UAEMex 4974/2020. We thank Miss Paulina Vazquez for the figures designing and improvement.

REFERENCES

[1]   Reynal, S.M.; Broderick, G.A.; Bearzi, C. Comparison of four markers for quantifying microbial protein flow from the rumen of lactating dairy cows. *J. Dairy Sci.,* **2005**, *88*(11), 4065-4082.
      [http://dx.doi.org/10.3168/jds.S0022-0302(05)73091-5] [PMID: 16230711]

[2]   Mariz, L.D.S.; Amaral, P.M.; Valadares Filho, S.C.; Santos, S.A.; Marcondes, M.I.; Prados, L.F.; Carneiro Pacheco, M.V.; Zanetti, D.; de Castro Menezes, G.C.; Faciola, A.P. Dietary protein reduction on microbial protein, amino acids digestibility, and body retention in beef cattle. I. Digestibility sites and ruminal synthesis estimated by purine bases and ^{15}N as markers. *J. Anim. Sci.,* **2018**, *96*(6), 2453-2467.
      [http://dx.doi.org/10.1093/jas/sky134] [PMID: 29668924]

[3]   Menezes, A.C.B.; Valadares Filho, S.C.; Costa e Silva, L.F.; Pacheco, M.V.C.; Pereira, J.M.V.; Rotta, P.P.; Zanetti, D.; Detmann, E.; Silva, F.A.S.; Godoi, L.A.; Rennó, L.N. Does a reduction in dietary crude protein content affect performance, nutrient requirements, nitrogen losses, and methane emissions in finishing Nellore bulls? *Agric. Ecosyst. Environ.,* **2016**, *223*, 239-249.
      [http://dx.doi.org/10.1016/j.agee.2016.03.015]

[4]   Ipharraguerre, I.R.; Reynal, S.M.; Liñeiro, M.; Broderick, G.A.; Clark, J.H. A comparison of sampling sites, digesta and microbial markers, and microbial references for assessing the postruminal supply of nutrients in dairy cows. *J. Dairy Sci.,* **2007**, *90*(4), 1904-1919.
      [http://dx.doi.org/10.3168/jds.2006-159] [PMID: 17369231]

[5]   Calsamiglia, S.; Stern, M.D.; Firkins, J.L. Comparison of nitrogen-15 and purines as microbial markers in continuous culture. *J. Anim. Sci.,* **1996**, *74*(6), 1375-1381.
      [http://dx.doi.org/10.2527/1996.7461375x] [PMID: 8791211]

[6]   Del Valle, T.A.; Ghizzi, L.G.; Zilio, E.M.C.; Marques, J.A.; Dias, M.S.S.; Silva, T.B.P.; Gheller, L.S.; Silva, G.G.; Sconamiglio, N.T.; Nunes, A.T.; Rennó, L.N.; Costa, V.E.; Rennó, F.P. Evaluation of ^{15}N and purine bases as microbial markers to estimate ruminal bacterial nitrogen outflows in dairy cows. *Anim. Feed Sci. Technol.,* **2019**, *258*114297
[http://dx.doi.org/10.1016/j.anifeedsci.2019.114297]

[7]   Hristov, A.N.; Bannink, A.; Crompton, L.A.; Huhtanen, P.; Kreuzer, M.; McGee, M.; Nozière, P.; Reynolds, C.K.; Bayat, A.R.; Yáñez-Ruiz, D.R.; Dijkstra, J.; Kebreab, E.; Schwarm, A.; Shingfield, K.J.; Yu, Z. Invited review: Nitrogen in ruminant nutrition: A review of measurement techniques. *J. Dairy Sci.,* **2019**, *102*(7), 5811-5852.
[http://dx.doi.org/10.3168/jds.2018-15829] [PMID: 31030912]

[8]   Broderick, G.A. Abrams, S.M.,Rotz, C.A. *Ruminal In Vitro* Degradability of Protein in Alfalfa Harvested as Standing Forage or Baled Hay., **1992**, *75*(9), 2440-2446.

[9]   Carro, M.D.; Miller, E.L. Comparison of microbial markers (15 N and purine bases) and bacterial isolates for the estimation of rumen microbial protein synthesis. *Anim. Sci.,* **2002**, *75*(2), 315-321.
[http://dx.doi.org/10.1017/S1357729800053078]

[10]   Perez, J.F.; Balcells, J.; Guada, J.A.; Castrillo, C. Determination of rumen microbial-nitrogen production in sheep: a comparison of urinary purine excretion with methods using ^{15}N and purine bases as markers of microbial-nitrogen entering the duodenum. *Br. J. Nutr.,* **1996**, *75*(5), 699-709.
[http://dx.doi.org/10.1079/BJN19960174] [PMID: 8695597]

[11]   Nursoy, H.; Ronquillo, M.G.; Faciola, A.P.; Broderick, G.A. Lactation response to soybean meal and rumen-protected methionine supplementation of corn silage-based diets. *J. Dairy Sci.,* **2018**, *101*(3), 2084-2095.
[http://dx.doi.org/10.3168/jds.2017-13227] [PMID: 29290449]

[12]   Nutritive requirements of ruminant animals. *Protein Nutrition Abstracts and Reviews Series B.,* **1992**, *62*, 787-835.

[13]   Moran, J. *How The Rumen Works in: Tropical Dairy Farming: Feeding Management for Small Holder Dairy Farmers in the Humid Tropics*; Landlinks Press, **2005**, pp. 42-49.
[http://dx.doi.org/10.1071/9780643093133]

[14]   Pacheco, D.; Waghorn, G.C. Dietary Nitrogen – definitions, digestión, excretion and consequences of excess for grazing ruminants. *Proceedings of the New Zealand Grassland Association.,* **2008**, *70*, 107-116.

[15]   Lapierre, H.; Pacheco, D.; Berthiaume, R.; Ouellet, D.R.; Schwab, C.G.; Dubreuil, P.; Holtrop, G.; Lobley, G.E. What is the true supply of amino acids for a dairy cow? *J. Dairy Sci.,* **2006**, *89* Suppl. 1, E1-E14.
[http://dx.doi.org/10.3168/jds.S0022-0302(06)72359-1] [PMID: 16527873]

[16]   Marini, J.C.; Fox, D.G.; Murphy, M.R. Nitrogen transactions along the gastrointestinal tract of cattle: A meta-analytical approach. *J. Anim. Sci.,* **2008**, *86*(3), 660-679.
[http://dx.doi.org/10.2527/jas.2007-0039] [PMID: 18073282]

[17]   Nozière, P.; Steinberg, W.; Silberberg, M.; Morgavi, D.P. Amylase addition increases starch ruminal digestion in first-lactation cows fed high and low starch diets. *J. Dairy Sci.,* **2014**, *97*(4), 2319-2328.
[http://dx.doi.org/10.3168/jds.2013-7095] [PMID: 24534508]

[18]   Sniffen, C.J.; O'Connor, J.D.; Van Soest, P.J.; Fox, D.G.; Russell, J.B. A net carbohydrate and protein system for evaluating cattle diets: II. Carbohydrate and protein availability. *J. Anim. Sci.,* **1992**, *70*(11), 3562-3577.
[http://dx.doi.org/10.2527/1992.70113562x] [PMID: 1459919]

[19]   Russell, J.B.; O'Connor, J.D.; Fox, D.G.; Van Soest, P.J.; Sniffen, C.J. A net carbohydrate and protein system for evaluating cattle diets: I. Ruminal fermentation. *J. Anim. Sci.,* **1992**, *70*(11), 3551-3561.
[http://dx.doi.org/10.2527/1992.70113551x] [PMID: 1459918]

[20]   Fox, D.G.; Sniffen, C.J.; O'Connor, J.D.; Russell, J.B.; Van Soest, P.J. A net carbohydrate and protein system for evaluating cattle diets: III. Cattle requirements and diet adequacy. *J. Anim. Sci.,* **1992**, *70*(11), 3578-3596.
[http://dx.doi.org/10.2527/1992.70113578x] [PMID: 1334063]

[21]   O'Connor, J.D.; Sniffen, C.J.; Fox, D.G.; Chalupa, W. A net carbohydrate and protein system for evaluating cattle diets: IV. Predicting amino acid adequacy. *J. Anim. Sci.,* **1993**, *71*(5), 1298-1311.
[http://dx.doi.org/10.2527/1993.7151298x] [PMID: 8505261]

[22]   Licitra, G.; Hernandez, T.M.; Van Soest, P.J. Standardization of procedures for nitrogen fractionation of ruminant feeds. *Anim. Feed Sci. Technol.,* **1996**, *57*, 347-358.
[http://dx.doi.org/10.1016/0377-8401(95)00837-3]

[23]   Broderick, G.A.; Merchen, N.R. Markers for quantifying microbial protein synthesis in the rumen. *J. Dairy Sci.,* **1992**, *75*(9), 2618-2632.
[http://dx.doi.org/10.3168/jds.S0022-0302(92)78024-2] [PMID: 1452862]

[24]   Couto Filho, C.C.C. SimõesSaliba, E.O., Rodriguez, N.M., Carneiro, Barbosa, G.S.S., Martins de Freitas, R., DinizResende, M.P. 2,6-Diaminopimelic acid (DAPA) in microbial protein quantification of heifers fed different forage sources. *Rev. Bras. Zootec.,* **2015**, *44*(5), 180-186.
[http://dx.doi.org/10.1590/S1806-92902015000500003]

[25]   Dewhurst, R. J.; Davies, D. R.; Merry, R. J. *Microbial protein supply from the rumen.,* **2000**.
[http://dx.doi.org/10.1016/S0377-8401(00)00139-5]

[26]   González Ronquillo, M. Balcells, J., Guada, J.A., and Vicente. F. Purine derivatives excretion in dairy cows: Endogenous excretion and the effect of exogenous nucleic acid supply. *J. Dairy Sci.,* **2003**, *86*, 1282-1291.
[http://dx.doi.org/10.3168/jds.S0022-0302(03)73712-6] [PMID: 12741553]

[27]   Carro, M.D. La determinación de la síntesis de proteína microbiana en el rumen: Comparación entre marcadores microbianos (Revisión). *Inv. Agr. Prod. Sanid. Anim.,* **2001**, *16*, 1-27.

[28]   Clark, J.H.; Klusmeyer, T.H.; Cameron, M.R. Microbial protein synthesis and flows of nitrogen fractions to the duodenum of dairy cows. *J. Dairy Sci.,* **1992**, *75*(8), 2304-2323.
[http://dx.doi.org/10.3168/jds.S0022-0302(92)77992-2] [PMID: 1401380]

[29]   Pérez, J.F.; Balcells, J.; Guada, J.A.; Castrillo, C. Rumen microbial production estimated either from urinary purine derivative excretion or from direct measurements of 15 N and purine bases as microbial markers: effect of protein source and rumen bacteria isolates. *Anim. Sci.,* **1997**, *65*(2), 225-236.
[http://dx.doi.org/10.1017/S1357729800016532]

[30]   Aharoni, Y.; Tagari, H. Use of nitrogen-15 determinations of purine nitrogen fraction of digesta to define nitrogen metabolism traits in the rumen. *J. Dairy Sci.,* **1991**, *74*(8), 2540-2547.
[http://dx.doi.org/10.3168/jds.S0022-0302(91)78431-2] [PMID: 1918533]

[31]   Hristov, A.N.; Huhtanen, P.; Rode, L.M.; Acharya, S.N.; McAllister, T.A. Comparison of the ruminal metabolism of nitrogen from ^{15}N-labeled alfalfa preserved as hay or as silage. *J. Dairy Sci.,* **2001**, *84*(12), 2738-2750.
[http://dx.doi.org/10.3168/jds.S0022-0302(01)74728-5] [PMID: 11814030]

[32]   Vaga, M.; Huhtanen, P. *In vitro* investigation of the ruminal digestion kinetics of different nitrogen fractions of ^{15}N-labelled timothy forage. *PLoS One,* **2018**, *13*(9)e0203385
[http://dx.doi.org/10.1371/journal.pone.0203385] [PMID: 30222744]

[33]   Ahvenjärvi, S.; Vaga, M.; Vanhatalo, A.; Huhtanen, P. Ruminal metabolism of grass silage soluble nitrogen fractions. *J. Dairy Sci.,* **2018**, *101*(1), 279-294.
[http://dx.doi.org/10.3168/jds.2016-12316] [PMID: 29103707]

[34]   Pérez, J.F.; Rodriguez, C.A.; Gonzalez, J.; Balcells, J.; Guada, J.A. Contribution of dietary purine bases to duodenal digesta in sheep. In situ studies of purine degradability corrected for microbial contamination. *Anim. Feed Sci. Technol.,* **1996**, *62*(2-4), 251-262.

[http://dx.doi.org/10.1016/S0377-8401(96)00961-3]

[35] Martın-Orúe, S.M.; Balcells, J.; Zakraoui, F.; Castrillo, C. Quantification and chemical composition of mixed bacteria harvested from solid fractions of rumen digesta: effect of detachment procedure. *Anim. Feed Sci. Technol.,* **1998**, *71*(3-4), 269-282.
[http://dx.doi.org/10.1016/S0377-8401(97)00156-9]

[36] Reynal, S.M.; Ipharraguerre, I.R.; Liñeiro, M.; Brito, A.F.; Broderick, G.A.; Clark, J.H. Omasal flow of soluble proteins, peptides, and free amino acids in dairy cows fed diets supplemented with proteins of varying ruminal degradabilities. *J. Dairy Sci.,* **2007**, *90*(4), 1887-1903.
[http://dx.doi.org/10.3168/jds.2006-158] [PMID: 17369230]

[37] Huhtanen, P.; Brotz, P.G.; Satter, L.D. Omasal sampling technique for assessing fermentative digestion in the forestomach of dairy cows. *J. Anim. Sci.,* **1997**, *75*(5), 1380-1392.
[http://dx.doi.org/10.2527/1997.7551380x] [PMID: 9159288]

[38] Ahvenjärvi, S.; Vanhatalo, A.; Huhtanen, P.; Varvikko, T. Determination of reticulo-rumen and whole-stomach digestion in lactating cows by omasal canal or duodenal sampling. *Br. J. Nutr.,* **2000**, *83*(1), 67-77.
[http://dx.doi.org/10.1017/S0007114500000106] [PMID: 10703466]

[39] Huhtanen, P.; Kaustell, K.; Jaakkola, S. The use of internal markers to predict total digestibility and duodenal flow of nutrients in cattle given 6 different diets. *Anim. Feed Sci. Technol.,* **1994**, *48*, 211-227.
[http://dx.doi.org/10.1016/0377-8401(94)90173-2]

[40] Siddons, R.C.; Paradine, J.; Beever, D.E.; Cornell, P.R. Ytterbium acetate as a particulate-phase digesta-flow marker. *Br. J. Nutr.,* **1985**, *54*(2), 509-519.
[http://dx.doi.org/10.1079/BJN19850136] [PMID: 2998454]

[41] Udén, P.; Colucci, P.E.; Van Soest, P.J. Investigation of chromium, cerium and cobalt as markers in digesta. Rate of passage studies. *J. Sci. Food Agric.,* **1980**, *31*(7), 625-632.
[http://dx.doi.org/10.1002/jsfa.2740310702] [PMID: 6779056]

[42] Armentano, L.E.; Russell, R.W. Method for calculating digesta flow and apparent absorption of nutrients from nonrepresentative samples of digesta. *J. Dairy Sci.,* **1985**, *68*(11), 3067-3070.
[http://dx.doi.org/10.3168/jds.S0022-0302(85)81204-2] [PMID: 4078133]

[43] France, J.; Siddons, R.C. Determination of digesta flow by continuous marker infusion. *J. Theor. Biol.,* **1986**, *121*, 105-120.
[http://dx.doi.org/10.1016/S0022-5193(86)80031-5]

[44] Brito, A.F.; Broderick, G.A.; Reynal, S.M. Effects of different protein supplements on omasal nutrient flow and microbial protein synthesis in lactating dairy cows. *J. Dairy Sci.,* **2007**, *90*(4), 1828-1841.
[http://dx.doi.org/10.3168/jds.2006-559] [PMID: 17369224]

[45] *Association of Analytical Chemists. Official Methods of Analysis,* 23rd ed; AOAC: Wahington, D.C., **2003**.

[46] Van Soest, P.J.; Robertson, J.B.; Lewis, B.A. Methods for dietary fiber, neutral detergent fiber, and nonstarch polysaccharides in relation to animal nutrition. *J. Dairy Sci.,* **1991**, *74*(10), 3583-3597.
[http://dx.doi.org/10.3168/jds.S0022-0302(91)78551-2] [PMID: 1660498]

[47] Hintz, R.W.; Mertens, D.R.; Albrecht, K.A. Effects of sodium sulfite on recovery and composition of detergent fiber and lignin. *J. AOAC Int.,* **1996**, *79*(1), 16-22.
[http://dx.doi.org/10.1093/jaoac/79.1.16] [PMID: 8620104]

[48] Broderick, G.A.; Huhtanen, P.; Ahvenjärvi, S.; Reynal, S.M.; Shingfield, K.J. Quantifying ruminal nitrogen metabolism using the omasal sampling technique in cattle--a meta-analysis. *J. Dairy Sci.,* **2010**, *93*(7), 3216-3230.
[http://dx.doi.org/10.3168/jds.2009-2989] [PMID: 20630238]

[49] Hristov, A.N. Comparative characterization of reticular and duodenal digesta and possibilities of

estimating microbial outflow from the rumen based on reticular sampling in dairy cows. *J. Anim. Sci.,* **2007**, *85*(10), 2606-2613.
[http://dx.doi.org/10.2527/jas.2006-852] [PMID: 17591704]

[50]   Krizsan, S.J.; Ahvenjärvi, S.; Volden, H.; Broderick, G.A. Estimation of rumen outflow in dairy cows fed grass silage-based diets by use of reticular sampling as an alternative to sampling from the omasal canal. *J. Dairy Sci.,* **2010**, *93*(3), 1138-1147.
[http://dx.doi.org/10.3168/jds.2009-2661] [PMID: 20172235]

[51]   Guzman-Cedillo, A.E.; Corona, L.; Castrejon-Pineda, F.; Rosiles-Martínez, R.; Gonzalez-Ronquillo, M. Evaluation of chromium oxide and titanium dioxide as inert markers for calculating apparent digestibility in sheep. *J. Appl. Anim. Res.,* **2017**, *45*, 275-279.
[http://dx.doi.org/10.1080/09712119.2016.1174124]

[52]   Reynal, S.M.; Broderick, G.A. Technical note: A new high-performance liquid chromatography purine assay for quantifying microbial flow. *J. Dairy Sci.,* **2009**, *92*(3), 1177-1181.
[http://dx.doi.org/10.3168/jds.2008-1479] [PMID: 19233811]

[53]   Komisarczuk, S.; Durand, M.; Beaumatin, P.H.; Hannequart, G. *Utilisation de l'azote 15 pour la mesure de la protéosynthèsemicrobiennedans les phases solide et liquide d'un fermenteur semi-continue*; Rusitec, **1987**.

[54]   Rodríguez, C.A.; González, J.; Alvir, M.R.; Repetto, J.L.; Centeno, C.; Lamrani, F. Composition of bacteria harvested from the liquid and solid fractions of the rumen of sheep as influenced by feed intake. *Br. J. Nutr.,* **2000**, *84*(3), 369-376.
[http://dx.doi.org/10.1017/S0007114500001653] [PMID: 10967616]

[55]   Hristov, A.N.; McAllister, T.A.; Ouellet, D.R.; Broderick, G.A. Comparison of purines and nitrogen-15 as microbial flow markers in beef heifers fed barley-or corn-based diets. *Can. J. Anim. Sci.,* **2005**, *85*(2), 211-222.
[http://dx.doi.org/10.4141/A04-054]

[56]   Carro, M.D.; Miller, E.L. Effect of supplementing a fibre basal diet with different nitrogen forms on ruminal fermentation and microbial growth in an *in vitro* semi-continuous culture system (RUSITEC). *Br. J. Nutr.,* **1999**, *82*(2), 149-157.
[http://dx.doi.org/10.1017/S0007114599001300] [PMID: 10743487]

[57]   Belenguer, A.; Yánez, D.; Balcells, J.; Ozdemir Baber, N.H.; González Ronquillo, M. Prediction of rumen microbial outflow urinary excretion of purine derivatives in milk-selected goats. *Livest. Prod. Sci.,* **2002**, *77*, 127-135.
[http://dx.doi.org/10.1016/S0301-6226(02)00081-7]

[58]   Liang, J.B.; Pimpa, O.; Abudalah, N.; Jelan, Z.A.; Nolan, J.V. Estimation of rumen microbial protein production from urinary purine derivatives in zebu cattle and wat buffalo., **1999**.

[59]   Verbic, J.; Chen, X.B.; MacLeod, N.A.; Orskov, E.R.Excretion of purine derivatives by ruminants. Effect of microbial nucleicacid infusion on purine derivative excretion by steer. *J. Agric.Sci. Camb,* **1990**, *114*, 243-248.
[http://dx.doi.org/10.1017/S0021859600072610]

[60]   Giesecke, D.; Stangassinger, M.; Tiemeyer, W. Nucleic acid digestion and urinary purines metabolites in sheep nourished by intra-gastric infusion. *Can. J. Anim. Sci.,* **1984**, *64*, 144-145.
[http://dx.doi.org/10.4141/cjas84-196]

[61]   Chen, X.B.; Ørskov, E.R.; Hovell, F.D. Excretion of purine derivatives by ruminants: endogenous excretion, differences between cattle and sheep. *Br. J. Nutr.,* **1990**, *63*(1), 121-129.
[http://dx.doi.org/10.1079/BJN19900097] [PMID: 2317473]

[62]   Balcells, J.; Guada, J.A.; Castrillo, C.; Gasa, J. Urinary excretion of allantoin and allantoin precursors by sheep after different rates of purine infusion into the duodenum. *J. Agri. Sci. Camb,* **1991**, *116*, 309-317.
[http://dx.doi.org/10.1017/S002185960007773X]

[63]   Giesecke, D.; Blasliemker, J.; Südekum, K.H.; Staingassinger, M. Plasma level, clearance and renal excretion of endogenous and ruminal purines in the bovine. *J. Anim. Physiol. Anim. Nutr. (Berl.),* **1993**, *70,* 180-189.
       [http://dx.doi.org/10.1111/j.1439-0396.1993.tb00321.x]

[64]   Balcells, J.; Guada, J.A.; Peiró, J.M.; Parker, D.S. Simultaneous determination of allantoin and oxypurines in biological fluids by high-performance liquid chromatography. *J. Chromatogr. A,* **1992**, *575*(1), 153-157.
       [http://dx.doi.org/10.1016/0378-4347(92)80517-T] [PMID: 1517293]

[65]   Faichney, G.J. *The use of markers to partition digestion within the gastrointestinal tract of ruminants. Digestion and Metabolism in the Ruminant*; MacDonald, I.W.; Warner, A.C.I., Eds.; University of New England Publishing Unit: Armidale, Australia, **1975**, pp. 277-291.

[66]   Martín Orúe, S.M.; Balcells, J.; Guada, J.A.; Castrillo, C. Endogenous purine and pyrimidine derivative excretion in pregnant sows. *Br. J. Nutr.,* **1995**, *73*(3), 375-385.
       [http://dx.doi.org/10.1079/BJN19950040] [PMID: 7766561]

[67]   Kerr, J.E.; Seraidarian, K. The separation of purine nucleosides from free purines and determination of the purines and ribose in these fractions. *J. Biol. Chem.,* **1945**, *159,* 211-225.

[68]   Young, E.G.; Conway, C.F. On the estimation of allantoin by the Rimini-Schryver reaction. *J. Biol. Chem.,* **1942**, *142,* 839-853.

CHAPTER 2

Qualitative and Quantitative LC-MS Analysis in Food Proteins and Peptides

Yi-Shen Zhu[1,*], Zhonghong Li[2] and Lulu Zhao[1]

[1] College of Biotechnology of Pharmaceutical Engineering, Nanjing Tech University, Nanjing211816, China

[2] Jiangsu Institute for Food and Drug Control, Nanjing210019, China

Abstract: LC-MS combines high separation ability of liquid chromatography with strong mass spectrometric structure identification. The advantages of LC-MS include high sensitivity and selectivity, minimal sample throughput, fast analysis speed and extensive structural information. It has been widely used in many fields, such as natural product analysis, pharmaceutical and food analysis, and environmental analysis.

In recent years, a great deal of researches have been conducted on the qualitative and quantitative aspects of food proteins and peptides. A variety of qualitative analyses of food proteins and peptides have been performed by LC-MS, such as accurate analysis of relative molecular weight, primary structural sequence, disulfide bond position, post-translational modifications (PTMs), *etc*. The quantitative analysis of proteins and peptides by LC-MS has been mainly achieved by two methods, *i.e.*, label-free methods (peak intensities approach and spectral counting approach) and labeled methods (chemical labeling, metabolic labeling and enzymatic labeling methods). This chapter focuses on the application of qualitative and quantitative analysis of proteins and peptides in food sources.

Keywords: Food proteins and peptides, Mass spectrometry, Qualitative analysis, Quantitative analysis.

INTRODUCTION

The rapid growth of proteomics research initiatives depends on the use of mass spectrometry (MS). With the development of soft ionization technology, the ability of MS to analyze proteins and peptides has been greatly improved. In 1988, Tanaka *et al*. [1] invented matrix-assisted laser desorption ionization (MALDI), where samples were dissolved in a suitable solvent and mixed with appropriate matrix, and ionized by laser beams. It produces singly charged species

---

[*] **Corresponding author Yi-Shen Zhu:** College of Biotechnology of Pharmaceutical Engineering, Nanjing Tech University, Nanjing, 211816, China; Tel: +86-18900660563; E-mail: zhuyish@njtech.edu.cn

Atta-ur-Rahman, M. Iqbal Choudhary & Syed Ghulam Musharraf (Eds.)

for biomacromolecules, which is very important for identifying molecular ions of proteins. In 1989, Fenn [2] proposed electrospray ionization (ESI) technology, where samples were dissolved in a buffer or polar solvent, and then introduced into the mass spectrometer in the form of a spray. It produces multiply charged species, which make the mass range of ESI theoretically unrestricted. At present, ESI-MS is usually used for qualitative and quantitative studies of various non-volatile and thermally labile samples [3]. These ionization technologies have greatly facilitated the widespread use of MS in proteomics. Fenn and Tanaka shared the 2002 Nobel Prize in Chemistry for their contribution on ESI and MALDI, respectively. Since then, MS developed rapidly as one of the most popular methods in analysis.

Various liquid chromatography (LC) and MS platforms have been organically integrated and are playing an increasingly important role in the field of proteins and peptides analysis. This chapter focuses on the researches and applications of LC-MS in the qualitative and quantitative aspects of food proteins and peptides.

QUALITATIVE MEHTODS OF LC-MS ON FOOD PROTEINS AND PEPTIDES

Induced by light, humidity or temperature, protein and peptide components in food may cause spatial changes due to subtle changes in the primary structure, post-translational modifications (PTMs) and disulfide bonds, leading to losses of nutrients. Qualitative characterization is particularly important for discovering the changes, which is the first step to avoid these losses.

Due to its high sensitivity and accuracy, LC-MS has been developed as a routine method for characterizing complex food protein mixtures, such as the analysis of precise relative molecular mass, primary structural sequences, disulfide bond positions, and PTMs, *etc.*

The first step is to digest sample proteins into peptides using proteolytic enzymes such as trypsin, chymotrypsin. After separation using one- or multidimensional LC, the digested peptides are ionized and the selected ions are sequenced to produce signature tandem mass spectrometry (MS/MS) spectra. Digested peptides are identified by using automated database search programs, which correlate the experimental MS/MS spectra with theoretical spectra predicted for each peptide contained in protein sequence database [4]. Several MS/MS database search tools are currently available, including SEQUEST [4a] and Mascot [4b] as widely used commercial applications, X!Tandem [4e], OMSSA [4f], and ProbID [5] as open source database search tools, and SpectrumMill [5] and Phenyx [6] as integrated programs (that provide other functionalities in addition to database search). Then peptide assignments are statistically validated and incorrect identifications are filtered out. Sequences of identified peptides are used to infer which proteins are present in the digested samples. Some peptides are presented in more than one protein, which complicates the protein inference process.

The Primary Structure and Relative Molecular Mass of Proteins and Peptides

Kou *et al*. [7] isolated and purified an antioxidant peptide CPe-III from the hydrolysates of chickpea albumin. Chickpea albumin isolates (CAI) were extracted from chickpea [8]. Chickpea albumin hydrolysates (CAH) were prepared from CAI by alcalase and flavorzyme proteases [9]. CAH was purified by gel filtration of Sephadex G-25, and lyophilized. The results of DPPH radical scavenging activity [10] showed that three peptide fractions obtained from gel filtration of CAH, Fraction I, Fraction II and Fraction III, were 23.15%, 34.02% and 41.3%, respectively. The results of hydroxyl radical scavenging activity [11] showed that at concentrations of 0.5 mg/mL, Fraction I, Fraction II and Fraction III inhibited hydroxyl radicals as 40.64%, 56.54% and 74.56%, respectively. The results of ABTS radical cation (ABTS*$^+$) decoloration assay [12] showed that Fraction III exhibited the highest antioxidative activity (0.967 ± 0.018 mmol/L). The reducing power [13] of Fraction III was the highest and exhibited a concentration dependence, as reported before [14].

Antioxidative peptides from Fraction III of CAH were purified and identified by RP-HPLC-ESI-MS/MS (LCQ Advantage MAX). From full scan LC-MS data, the highest peak was identified as a decapeptide (RQSHFANAQP, mw 1155 Da), corresponding to CPe-III, by Xcalibur software and BioWorks 3.3. CPe-III could be a suitable natural antioxidant in chickpea-related food.

Zhang's team [15] isolated and identified antioxidative peptides from rice endosperm protein (REP) enzymatic hydrolysate. First, REP was defatted and freeze-dried. Then, REP was hydrolysed with alcalase, chymotrypsin, papain, flavorase and neutrase, respectively, followed by measuring the antioxidant activities. The results of radical scavenging activities showed that these five enzymatic hydrolysates had the ability to scavenge DPPH radicals. At a concentration of 1.50 mg/ml, the values were 71.26 ± 1.06%, 35.26 ± 1.34%, 44.31 ± 2.12%, 24.31 ± 1.66%, and 85.86 ± 1.33% corresponding to alcalase, chymotrypsin, papain, flavorase and neutrase, respectively. Moreover, the results of superoxide radical (O_2^-) and hydroxyl radical (˙OH) scavenging capacities suggested that at 2.00 mg/ml, the neutrase hydrolysate had the highest values, 75.69±1.23% and 82.93±1.21%, respectively. Fraction 3 of the neutrase hydrolysate had the highest antioxidant activity in four ion-exchange isolated fractions. The subfractions F3b and F3c further separated by RP-HPLC from

Fraction 3 possessed higher antioxidant activities. MALDI-TOF/TOF MS/MS analysis determined the molecular mass of F3b and F3c were 959.5 Da and 1002.5 Da, respectively. Direct amino acid sequencing was failed by Biolynx peptide sequencer. Manual calculation and amino acid composition analysis presented F3b as FRDEKK and F3c as KHNRGDEF. FRDEHKK was synthesized and the antioxidant activity was confirmed in the linoleic acid model system and radical-induced cell culture assay. It was feasible to produce natural antioxidants from REP.

Corn peptides (CP) from zein hydrolysates digested by immobilized alcalase and trypsin were isolated and identified [16]. A novel antioxidant and antihypertensive peptide CP-2-1 was obtained from HPLC isolation of CP-2 fraction. The amino acid sequence measured by LC-MS/MS (micrOTOF-Q II) was M-(I/L)-P-P with a molecular weight of 452.3Da, and the antioxidant activity and angiotensin I-converting enzyme (ACE) inhibitory activity were IC_{50} 220 µg/mL and 70.32 µg/mL, respectively. It holds promise as a leading compound for antioxidant functional foods or antihypertensive drugs.

Intramolecular Disulfide Bonds Analysis of Proteins

Nagao *et al.* [17] determined four disulfide bonds by LC-MS/MS from a peptide called La1, which consisted of 73 amino acid residues (Fig. **1**), from the scorpion *Liocheles australasiae* venom. La1, which was chemically synthesized using a natural chemical ligation (NCL) method, was digested by trypsin. Without a reduction in disulfide bonds, these tryptic digestions of La1 were analyzed by LC-MS. A tryptic digested peptide, which was cross-linked by a disulfide bond between Cys5 and Cys29, was detected (Retention time 18.1 min). And another tryptic digested peptide, which was cross-linked by other three disulfide bonds, was also detected (Retention time 20.3 min). After HPLC purification, the peptide with three disulfide bonds was cleaved at Met69 by CNBr. MALDI-TOF MS analysis demonstrated that a cleaved peptide (m/z 841.4) corresponding to SGCK and VTCK linked by a disulfide bond between Cys52 and Cys72, and another cleaved peptide (m/z 2872.3) corresponding to TVPGGAGAAFPSCCPh, CATVNLK, and DCALYK linked by other two disulfide bridges. Further analysis of MALDI-TOF/TOF MS was assigned on unknown disulfide bond linkages (Table **1**). Two major product ions (m/z 2044.0 and 2672.0) with fragment ions generated by cleavage between Cys66 and Cys67 were observed, which allowed the distinction between two possible linkages of disulfide bonds. A fragment (m/z 1047.5) corresponding to CATVNLK and CPh linked with a disulfide bond between Cys43 and Cys67 was observed. The fragment ions (m/z 997.3) confirmed DCALYK and PSC linked with another disulfide bond between Cys24 and Cys66. Moreover, the absence of fragment ions containing a disulfide bond between Cys43 and Cys66 (m/z 1033.5) or between Cys24 and Cys67 (m/z 1011.4) further supported the identified disulfide bonds.

Structural Analysis of PTMs Proteins and Peptides

PTMs play critical roles in protein functional regulation [18]. About 200 types of PTMs have been reported so far, such as phosphorylation [19], glycosylation [20], methylation [21] and acetylation [22].

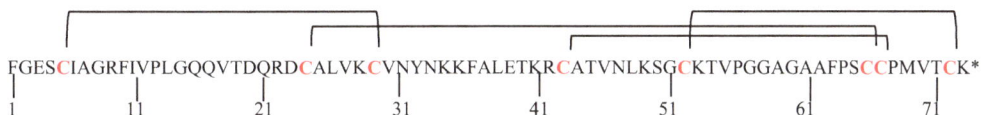

Fig. (1). Amino acid sequence and disulfide bonding pattern of La1. The asterisk is the amidated c-terminal.

Table 1. MS/MS analysis of the peptide fragment containing two disulfide bonds.

No.	Sequence	Mass-to-charge Ratio(m/z)
1	PS^{66}C D^{24}CALYK	997.3
2	^{43}CATVNLK ^{67}CPh	1047.5
3	^{43}CATVNLK PS^{66}C^{67}CPh D^{24}CALYK	2044.0
4	^{43}CATVNLK PGGAGAAFPS^{66}C^{67}CPh D^{24}CALYK	2672.0

The casein phosphorylated peptides (CPPs) were analyzed by LC-ESI-MS, and the stability of serine phosphorylation on β-CN f1-25 4P were investigated during the hydrolysis of CPPs [23]. The hydrolysates were digested from reconstituted sodium casein (NaCN) by TPCK-treated trypsin in 0.5, 1.0, 1.5, 2.0 and 3.0 hours. The supernatant of digestions was heat-treated. Calcium chloride and ethanol were added subsequentially for phosphopeptide enrichment [24]. The preparation and enrichment scheme of CPPs is shown in Fig. (**2**). The precipitates were resolubilised and analyzed by LC-ESI-MS and MS/MS. All the MS/MS spectra were analyzed using Data Analysis software (version 4.0, Bruker Daltonics) and BioTool (version 3.1, Bruker Daltonics). Data were searched on the MASCOT (Matrix Science, v2.204) search engine against the ExPasy protein database (SwissProt v51.6) using variable modifications: phosphorylation of serine/threonine. Incomplete post translational serine phosphorylation *in vivo* or dephosphorylation during NaCN manufacture may explain the different levels of phosphorylation observed for β-CN f1-25 4P (Fig. **3**). It was the first time that the presence of all these phosphorylated forms of β-CN f1-25 had been reported in a single NaCN hydrolysate. The results showed that the stability of pH 7 serine phosphorylation was better at the test temperature of 75 °C.

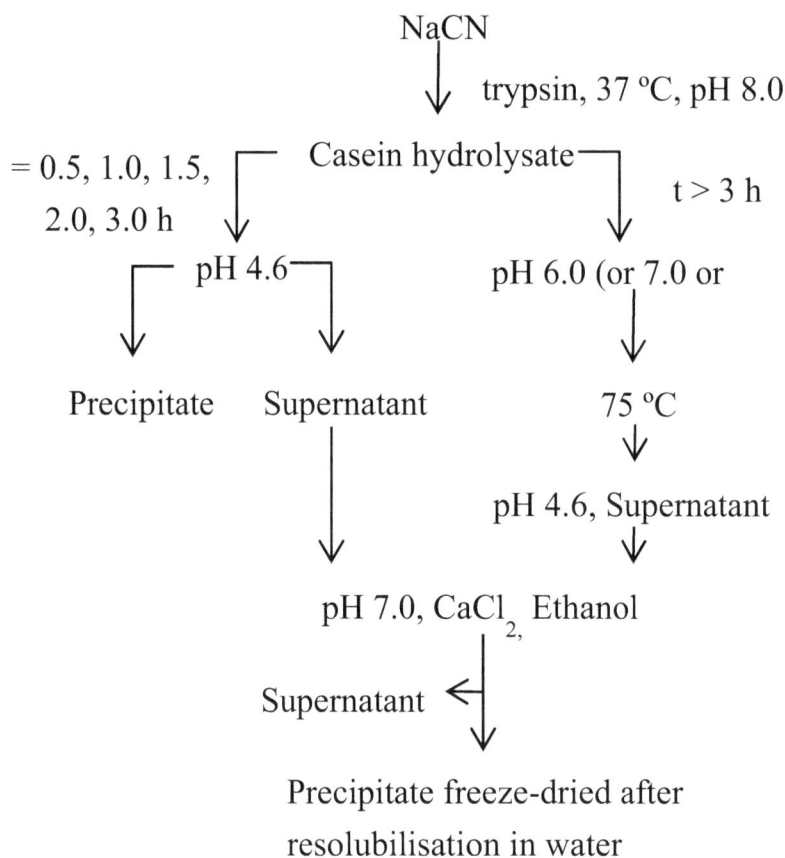

NaCN

↓ trypsin, 37 °C, pH 8.0

┌── Casein hydrolysate ──┐

= 0.5, 1.0, 1.5,   t > 3 h
2.0, 3.0 h   ↓

┌── pH 4.6 ──┐   pH 6.0 (or 7.0 or

↓   ↓   ↓

Precipitate   Supernatant   75 °C

                  │   ↓

                  │   pH 4.6, Supernatant

                  │   ↓

pH 7.0, CaCl$_2$, Ethanol

Supernatant ⬅┤

↓

Precipitate freeze-dried after
resolubilisation in water

Fig. (2). Preparation and enrichment scheme of casein phosphorylated peptides.

Zhu *et al.* [25] further enriched CPPs with gallium IMAC, analyzed and identified CPPs by LC-ESI-MS. The results showed that gallium IMAC combined with calcium aggregation and ethanol precipitation was an effective method for CPPs enrichment. The results helped to establish the quality control methods of CPPs in industrial production.

A novel glycoside assay was developed by Hua *et al.* [26], which showed the glycan isomer differentiation at specific sites and identified at least 13 different glycans (including isomers). First, a non-specific protease was used to hydrolyze glycoproteins into glycopeptides. Then, the glycopeptides were analyzed by LC-ESI-MS. The results demonstrated an N-glycosylation site on bovine ribonuclease B, Asn at position 60, were linking 13 glycans; 5 N-glycosylation sites on bovine lactoferrin (Fig. **4a**), Asn at positions 252 (Fig. **4b**), 300, 387, 495 and 564, were linking 59 glycans; five O-glycosylation sites on the bovine κ-casein, Thr at positions 142, 152, 154, 157 and 163, were linking 20 glycans.

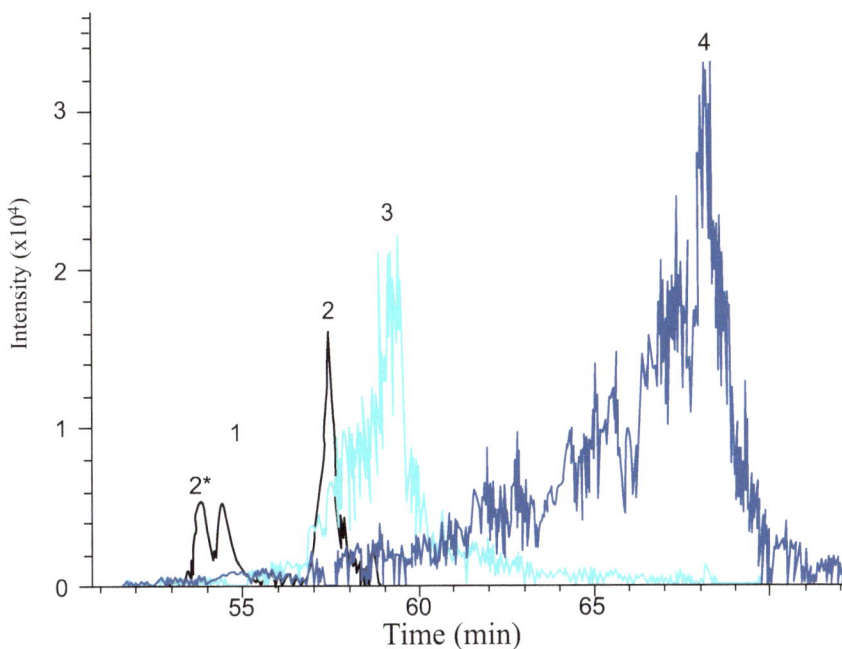

Fig. (3). The extracted intensity chromatogram (EIC) of multi-phosphorylated peptides corresponding to β-CN f1–25. (1) Mono-phosphorylated peptide. (2) Di-phosphorylated peptide with phosphoseryl residues at positions 18 and 19. (2') Di-phosphorylated peptide with phosphoseryl residues at positions 15 and 19. (3) Tri-phosphorylated peptide with phosphoseryl residues at positions 17, 18 and 19. (4) Tetra-phosphorylated peptide. Tandem mass spectrum showing extensive neutral loss of phosphoric acids for triply charged ions: m/z 961.450 (1), 988.104 (2), 988.111 (2'), 1014.761 (3) and 1041.404 (4), numbers in brackets and in the figure above correspond to the peaks in the relevant EICs. Reproduced with permission from [23].

LC-MS has a wide range of applications in the characterization of food proteins and peptides. However, qualitative analysis provides less information about protein quantitation, protein composition, and protein levels.

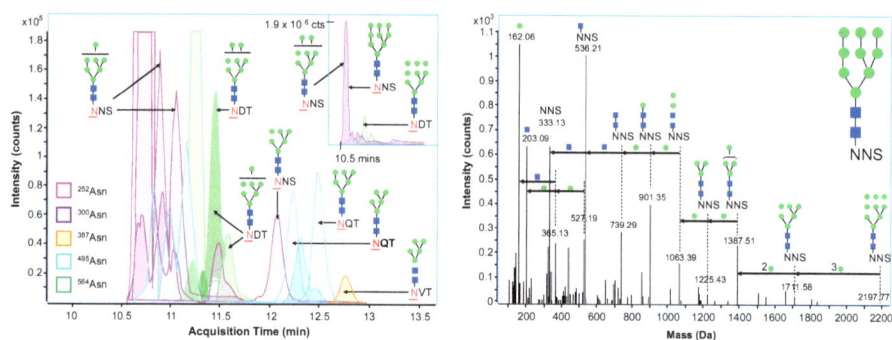

Fig. (4). (a) Overlaid chromatograms and associated structural assignments of glycopeptides from bovine lactoferrin. Inset, a zoomed-out view of the same chromatograms, showed the dynamic range of the glycopeptide mixture. Color denoted the site of glycosylation from which the glycopeptide originated. **(4b)** Deconvoluted tandem MS spectrum of a bovine lactoferrin glycopeptide. The presence of lone peptide as well as HexNAc-conjugated peptide fragment peaks identified [252]NNS + Man9 as the correct glycopeptide composition. Reproduced with permission from [26].

QUANTITATIVE METHODS OF LC-MS ON FOOD PROTEINS AND PEPTIDES

LC-MS is an analytical chemistry technique which includes the ability of separation by LC and the advantages of high resolution, high sensitivity and selectivity by MS. Furthermore, LC-MS is suitable for the analysis of samples with high molecular weight or thermal instability. Based on these advantages, LC-MS analysis plays an important role in quantitative proteomics. Quantitative proteomics can be classified into labeled quantitation and label-free quantitation methods.

Labeled Quantitation Methods

According to the coupling strategies, labeled quantitation methods can be divided into chemical labeling method, metabolic labeling method and enzymatic labeling method. Samples are differentially labeled and mixed, followed by LC-MS analysis. The mass difference of labels is used to identify the sample. The peak area ratio of labeled peptide segments in MS/MS is used to quantify [27] (Fig. **5**).

Chemical Labeling Method

According to the difference of the reactive groups, chemical labeling method can be classified into sulfhydryl labeling method, amino labeling method, and carboxyl labeling method [28]. It has been widely used in the field of quantitative proteomics.

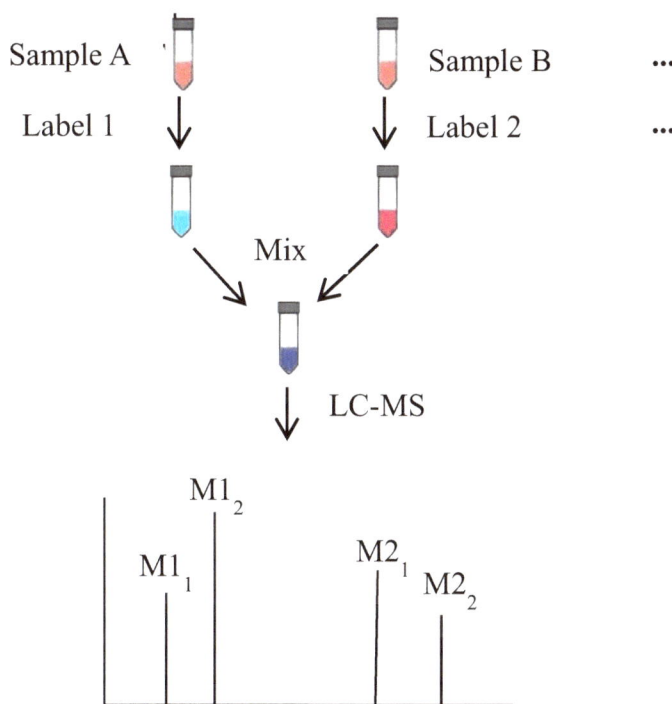

Fig. (5). General process for labeled quantitation methods.

Sulfhydryl Labeling Method

In 1999, Gygi *et al*. [29] first introduced the isotope coded affinity tags (ICAT) technology labeling cysteine residues in samples with light or heavy isotopes. The ICAT reagent consists of three parts. [1] an iodoacetyl reactive group, targeting cysteinyl residues; [2] a linker region consisting eight heavy (^{2}H) or light (^{1}H) atoms; and [3] a biotin group suitable for affinity purification [29]. One of the major pitfalls of this method was the difference in retention time in the light and heavy labeled peptides during LC caused by the ^{2}H atoms [30]. The ICAT reagent structure and reaction mechanism are shown in Fig. (**6**).

Prior to ICAT technology, the methods for identifying oxidants-sensitive cysteines (redox proteomics) were based on the reaction of iodoacetamide (IAM) with free cysteine thiols, and the fact that thiol oxidation prevented binding [31]. The ICAT method is superior to previous redox proteomics methods because (1) does not require gel separation of proteins, and (2) simultaneously quantifies the degree of oxidation of the involved Cys residue and successfully identifies the peptide/protein [32].

Fig. (6). Structure and reaction mechanism of the isotope coded affinity tags (ICAT) reagent.

Application of ICAT Technology in Animal Protein Research

ICAT technology was used by Sethuraman *et al.* [32] to simultaneously identify and quantify oxidant-sensitive cysteine thiols that are sensitive to high concentrations of hydrogen peroxide in membrane particulate fractions (MPFs) from rabbit heart homogenate, and simultaneously identified known and some novel proteins with oxidant-sensitive cysteines.

The experimental process was briefly described as follows. MPFs were prepared from rabbit heart homogenate by sucrose gradient ultracentrifugation method. MPFs and hydrogen peroxide treated MPFs were incubated with light or heavy acid-cleavable ICAT reagent without reducing agent, respectively. After trypsin digestion, the digested peptides were separated by a strong cation exchange (SCX) column and an avidin affinity cartridge. The avidin affinity purified peptides were suspended in acid cleavage reagent to release the ICAT-labeled peptides. The peptides obtained by acid cleavage were suspended for μHPLC-ESI MS/MS analysis.

The results showed that glyceraldehyde-3-phosphate dehydrogenase (GAPDH), a key enzyme in glycolysis, was identified as a protein with known oxidant-sensitive cysteines. Some proteins with novel oxidant-sensitive Cys residues were also identified, such as phosphorylase B kinase beta regulatory chain, voltage-dependent anion selective channels 1, 2 and 3 (VDAC1,2 and 3), and hemoglobin β chain. 29 peptides with cysteines from 18 proteins were identified and quantified, 17% of the thiols in cysteine-containing peptides were oxidized by more than 50%, and only 10% were oxidized more than 70%.

Application of ICAT Technology in Plant Protein Research

Islam *et al.* [33] used ICAT technology to study the relationship between wheat chromosome deletion and protein expression. The genetic locus of wheat quality protein identified in this study can be used as markers in breeding strategies for high-quality wheat seeds.

The experiment was similar to the process in 3.1.1(1) (a) with slight modifications. The purified and desalted peptides were analyzed by µLC-ES--MS/MS. Data acquisition was performed by MassLynx and peptide sequence analysis was performed by BioLynx (Micromass). Sequence information was submitted to the SWISS-PROT and EMBL protein and peptide databases for identification.

With deletion at the end of 1B chromosome, there were 14 peptides quantified by mass fragment analysis of the euploid, in which 4 were down-regulated, 9 were up-regulated, and 1 was without change. Furthermore, four peptides were identified as α-amylase inhibitor, α-amylase/subtilisin inhibitor precursor, proteasome subunit α-type 7 and 1,4 α-glucan-D-maltohydrolase. It was the first time that the use of ICAT-ESI MS/MS avoided the complexity of wheat protein identification, and more sensitive and accurate separation of proteins and their expression. Unfortunately, it was often difficult to identify C-terminal fragments due to signals resulting from the fragmentation of ICAT reagents. And a major portion of the proteins remained insoluble in the ICAT-denaturing buffer which has limitation in making a useful correlation between the quantitative ICAT and 2-DE data.

Application of ICAT Technology in Microbial Protein Research

Flory and his partners [34] used second-generation $^{13}C/^{12}C$-based, acid-cleavable ICAT reagents to quantitatively profile cell cycle-related changes in proteomic expression in a model eukaryote, the budding yeast *S. cerevisiae*, which was the first time on a series of dynamic measurements for global eukaryotic proteomics using stable isotope labeling and MS/MS. These proteomic data sets provided a necessary and complementary measure of gene expression in eukaryotes, providing a database for functional analysis of *Saccharomyces cerevisiae* proteins and would also help to further develop the global proteomics analysis techniques for higher eukaryotes.

The experiment was similar the process in 3.1.1(a) with slight modifications. For protein identification, all MS/MS spectra were analyzed using SEQUEST, comparing experimental data with protein sequences in the *S. cerevisiae* genome database (SGD) [4a]. PeptideProphet (http://peptideprophet.sourceforge.net/) was used to assess the accuracy of the peptide assignments to MS/MS spectra made by SEQUEST [35]. Proteomic data were uploaded to the PeptideAtlas and SBEAMS (Systems Biology Experiment Analysis System) databases for proteomics data storage, sorting, and analysis, which are available *via* the platform to the public at: http://www.peptideatlas.org/repository/publications/flory2005/.

By using new ICAT reagents in combination with two dimensions of off-line chromatography and RPLC-MS/MS, the quantitative proteomic sampling of budding yeast was achieved with 2754 proteins or an estimated 48% of the *S. cerevisiae* predicted proteome.

Amino Labeling Method

The isobaric tags for relative and absolute quantitation (iTRAQ) technology is an *in vitro* isotope labeling technology introduced by Applied Biosystems Inc (ABI) in 2004, that can simultaneously conduct relative or absolute quantitation studies on up to eight protein samples [36]. Each iTRAQ reagent consists of three parts [36, 37]: (1) reporter group, consisting of label molecules with relative molecular masses of 113, 114, 115, 116, 117, 118, 119 and 121, respectively; (2) balance group, consisting of label molecules with relative molecular masses of 192, 191, 190, 189, 188, 187, 186, and 184, respectively; (3) reactive group, an identical peptide reactive group. Since the phenylalanine m/z is 120, there is no labeling molecule with a relative molecular mass of 120 in the reporter group to avoid the interference from phenylalanine [36]. The iTRAQ labeling structure is shown in Fig. (7).

Regardless of the iTRAQ reagent used, the total molecular weight of the reporter group and the balance group is 305. In MS results, peaks of the same peptides labeled with different iTRAQ reagents appear as same parent ion mass, which greatly reduces MS complexity. However, each iTRAQ result still produces thousands of spectra, and the amount of data is large and complex, which limits the application of iTRAQ technology. Therefore, tools and analytical methods of bioinformatics are essential for the analysis and interpretation of these data. Pro Quant and Protein Pilot provided by iTRAQ reagent manufacturers and other free software can be used to analyze these data [38].

The enzyme-digested peptides can be labeled to the N-terminal or each lysine side chain with iTRAQ reagents, which greatly improved the coverage and reliability of protein identification. Therefore, iTRAQ technology is suitable for all types of digested protein samples, such as low-abundance proteins, membrane proteins, partial base proteins, and hydrophobic proteins.

Balance group
Mass = 184~192(besides 185)

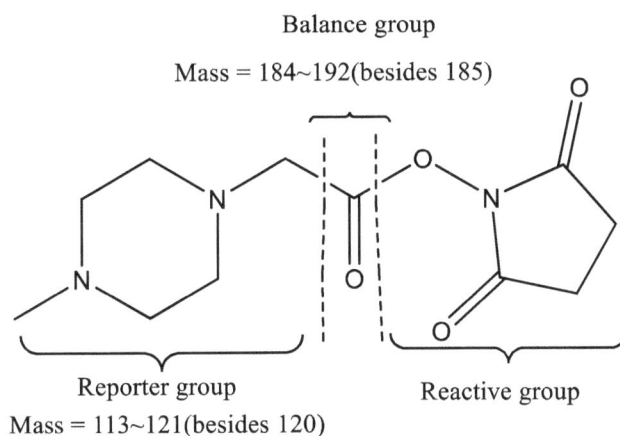

Reporter group
Mass = 113~121(besides 120)

Reactive group

Fig. (7). Structure of 8 different isotopic reagents.

Application of iTRAQ Technology in Animal Protein Research

Golovan and his cooperators [39] used iTRAQ technology to analyze *Sus scrofa* liver proteome and quantify proteins differentially expressed among genders, transgenic pigs and conventional pigs. This research generated a comprehensive protein catalog of pig liver proteins, which helped better understanding on the differences in liver proteomes in different animals, and provided a theoretical basis for improving pork quality.

Four different liver protein samples, a female and a male of conventional Yorkshire pigs (CF, CM) and transgenic line pigs (TF, TM), were extracted in each iTRAQ run. Three iTRAQ runs were carried out with 12 individual liver samples. The alkylated and digested protein samples were differentially labeled with four iTRAQ reagents with reporter groups of 114, 115, 116 and 117, respectively, followed by purifying on SCX column and analyzing by LC-ES--MS/MS. This experiment used the Paragon™ search algorithm of ProteinPilot™ v2.0 [40] to perform statistical analysis.

69,566 MS spectra from three iTRAQ runs were searched against the MSDB database leading to identification of 1476 proteins, from which 1123 proteins matched to mammalian species. When false positive (FP) identifications rate was set to less than 5%, a list of 880 liver proteins was confident. These proteins, which involved in energy metabolism, catabolism, protein biosynthesis, electron transport, and other oxidoreductase reactions, were highly enriched, confirming liver as a major chemical and energy plant. Sex and transgene were demonstrated to affect around 5% of total liver proteome, separately. When the Storey q value was used to control the false discovery rate (FDR) to within 0.1%, four proteins (EPHX1, CAT, PAH, ST13) and two proteins (SELENBP2, TAGLN) were differentially expressed on sex (male/female), and on lines (transgenic pigs/conventional pigs), respectively.

Liu and his colleagues [41] used iTRAQ technology to compare the protein expression differences during the growth of northeast Chinese sika deer antlers, and to screen functional proteins related to the growth of antler. There were 25 proteins significantly up-regulated and 21 proteins significantly down-regulated in the early stage of growth of two-branched antler (45 d), which were mainly related to cytoskeleton formation, tissue and organ development. When two-branched antler reached the end stage of growth (60 d), the tissue and cell anabolism were more vigorous, and a large number of functional proteins were secreted. The antler products harvested in this period were proved to have better quality with superior effects on clinical remedy.

iTRAQ technology was used to analyze bovine α-caseins' preparation digested with glutamyl endopeptidase (GE) at 37 and 50 °C by Zhu *et al*. [42]. For the identified α_{s1}-casein peptides and α_{s2}-casein peptides, the sequence coverages were 56% and 60%, and 26% and 30% at 37 and 50 °C, respectively. These data showed that incubation temperature had an important effect on promoting hydrolysis. Furthermore, large peptides were generated earlier than small sequences during GE hydrolysis of α-caseins. Non-phosphorylated peptides were more sensitive to LC-MS detection than phosphorylated peptides according to the MS intensities. However, there was no distinct difference in digestion rates between phosphorylated and non-phosphorylated peptides compared by the ratio of the iTRAQ ions. It was the first time to quantitatively demonstrate the substrate specificity of GE on α-caseins. The results helped to confirm the precise process of α-caseins' hydrolysis by GE, which was significant for quantifying the release of bio- and techno-functional peptides. Zhu *et al*. [43] also used iTRAQ technology to analyze bovine β-casein hydrolysates obtained using glutamyl endopeptidase (GE) at 37 and 50 °C. A quantitation methodology was setup to confirm the precise process of β-casein hydrolysis by GE, which was significant for quantifying the process of bioactive peptides from industry of food protein hydrolysate.

Application of iTRAQ Technology in Plant Protein Research

Zeng and his co-worker [44] used eight iTRAQ reagents to study the differential protein expression in natural selenium-enriched and non-selenium-enriched rice. This study not only provided interesting insights to proteomic analysis and to differential protein expression in selenium-enriched and non-selenium-enriched rice, but also provided basic information on protein mechanisms and secondary metabolite biases in the two rice varieties. It laid foundation for further research on the antioxidant and anti-aging mechanisms of selenium-enriched rice.

3235 proteins were identified and 3161 proteins were quantified, in which 401 were differential proteins. Among these 401 differential proteins, 208 were down-regulated and 193 were up-regulated. 77 significantly differentiated proteins were further analyzed and divided into 10 categories: oxidoreductase, transferase, isomerase, heat shock protein, lyase, hydrolase, ligase, synthase, tubulin and actin. The anti-stress, anti-oxidation, active oxygen metabolism, carbohydrate and amino acid metabolism in natural selenium-enriched rice were higher than those in non-selenium rice. The starch synthesis pathway of non-selenium-enriched rice was more active than that of selenium-enriched rice. Inferred from the Gene Ontology (GO), Clusters of Orthologous Groups (COG) and Kyoto Encyclopedia of Genes and Genomes (KEGG) annotations, cysteine synthetase (CYS) and methyltransferase (metE) may be two key proteins responsible for the difference in cysteine and carbohydrates, and OsAPx02, CatC, riPHGPX, HSP70 and HSP90 may be key enzymes that regulated the difference in antioxidant and anti-stress effects between the two rice varieties.

Chen and his colleagues [45] used iTRAQ technology to analyze black rice grain development, revealing metabolic pathways associated with anthocyanin (ACN) biosynthesis. Throughout 5 successive developmental stages, starting from 3 to 20 days after flowering (DAF), about 928 proteins (Fig. **8**) were identified, of which 230 were differentially expressed. The maximum number of differentially expressed proteins, 76 up-regulated proteins and 39 down-regulated proteins, was identified on 10/7 DAF. In contrast, the lowest number of differentially expressed proteins, 25 up-regulated proteins and 28 down-regulated proteins, was identified on 20/15 DAF (Fig. **9**). 230 differentially expressed proteins could be divided into 14 functional groups according to biological process analysis of GO, proteins in the largest group (26.09%) were involved in metabolic processes. Specifically, proteins playing roles in anthocyanin biosynthesis, sugar synthesis, and gene expression regulation were up-regulated at the onset of black rice grain development and especially during the development. In contrast, proteins involved in signal transduction, redox homeostasis, photosynthesis, and N-metabolism were down-regulated during black rice grain maturation. This experiment provided valuable new insights into the production of anthocyanins in black rice.

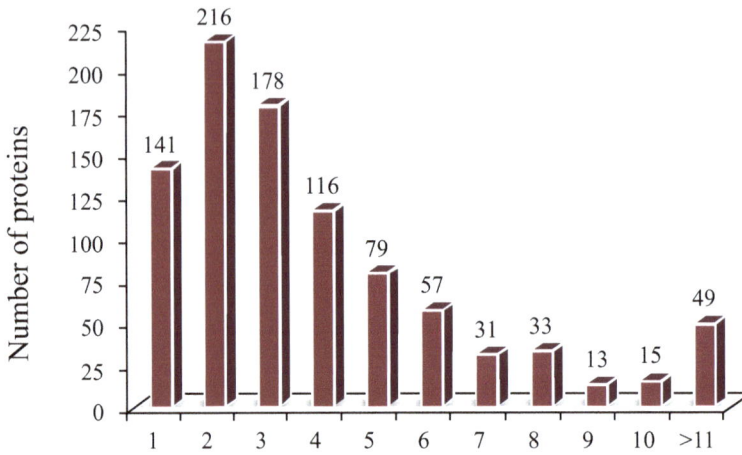

Fig. (8). Unique peptides of the identified proteins. Reproduced with permission from [45].

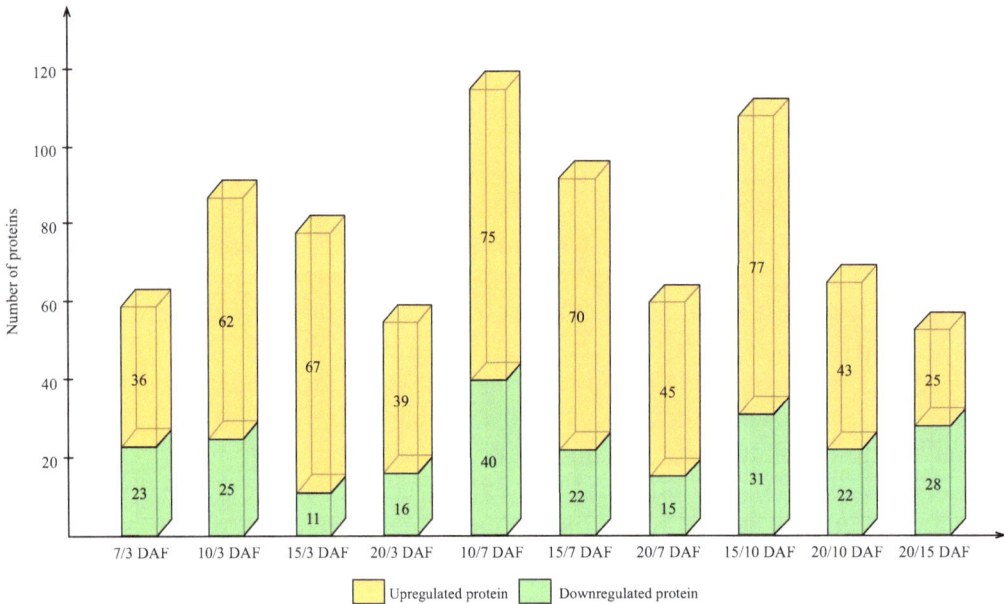

Fig. (9). The number of differentially expressed proteins in 5 development stages. The x-axis indicates the comparisons between any two samples. Reproduced with permission from [45].

iTRAQ technology is suitable for the study of plants under various stresses.

Yang *et al.* [46] first investigated molecular responses of plantain to cold stress by iTRAQ technology, which had potential value for developing cold tolerant banana cultivars. 3477 plantain proteins were identified, of which 809 differentially expressed proteins were quantified. The majority of differentially expressed proteins were involved in oxidation-reduction, including oxylipin biosynthesis, whereas others were associated with photosynthesis, photorespiration, and several primary metabolic processes, such as carbohydrate metabolic process and fatty acid β-oxidation.

Application of iTRAQ Technology in Fungal Protein Research

Ming and his partners [47] used iTRAQ technology to analyze the difference in free amino acids of *Russula griseocarnosa* harvested at different stages of maturity. The content of total free amino acid increased slightly from the first stage of 19.88 mg/g dry weight (DW) to the second stage of 20.57 mg/g DW, and then significantly decreased to 14.42 mg/g DW in the third stage. With the development of fruiting bodies, the monosodium glutamate (MSG)-like components such as Asp and Glu showed a downward trend. The contents of MSG-like components were in the range of 0.99-2.28 mg/g DW, which was lower than the sweet components (2.61-4.35 mg/g DW) or the bitter components (1.89-3.39 mg/g DW). The contents of the total tasteless components fluctuated in the range of 0.37 to 0.61 mg/g DW. Gln, Glu and Ala were the most important free amino acids of proteins and the quantitative differences of free amino acids became larger between individuals at the same stages of maturity. Further evidence was provided for the nutritional value and beneficial effects of the red mushroom.

iTRAQ technology combined with LC-MS/MS can simultaneously identify and quantify lots of proteins, and maximize the "full set of information" of the proteome. iTRAQ technology will play a more active role in future proteomics research.

Carboxyl Labeling Method

In 2001, Goodlett *et al*. [48] used methyl esterification of peptides to quantify proteins between two mixtures. This was the first report of carboxyl labeling method applied to the quantitative study of proteomics. Carboxyl labeling method can reduce experimental errors and improve the accuracy of relative quantitation [49]. A series of isotope-labeled reagents for carboxyl-based derivatization (Fig. **10**) have been designed and synthesized since then [50].

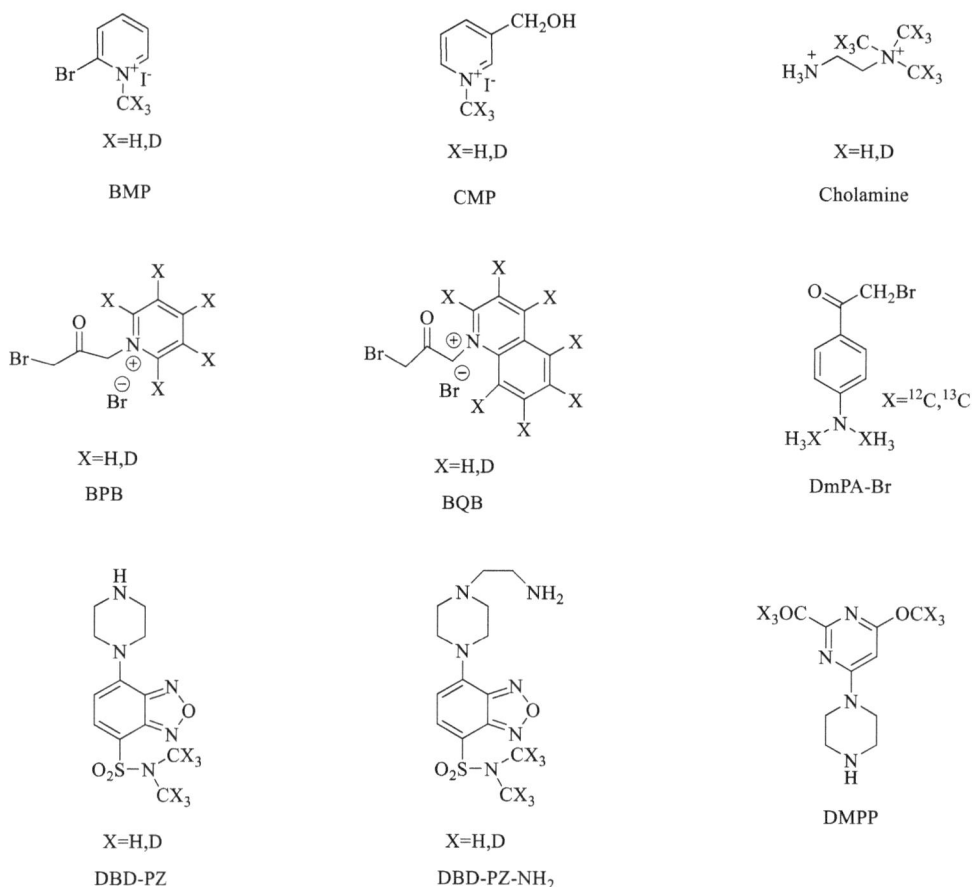

Fig. (10). The structures of carboxylic acid-derivate reagents.

Application of Carboxyl Labeling Method in Fatty Acids and Its Derivatives

Lamos *et al*. [51] synthesized a new labeling reagent, heavy isotope-labeled cholamine ($^2H_3$-cholamine), which was suitable for the relative quantitation of carboxylic acid-containing metabolites of small molecules and peptides. Chickens were fed diets differing in lipid composition and 12 fatty acids in the chicken eggs were simultaneously quantitated with this derivatization reaction.

2-bromo-1-methylpyridinium iodide (BMP) and 3-carbinol-1-methylpyridinium iodide (CMP) were developed by Yang *et al*. [52], which greatly improved ionization in positive-ion mode of ESI-MS [53]. After derivatization, MS detection sensitivity was significantly improved. Huang *et al*. [54] used chemically synthesized $^2H_5/H_5$-ω-bromoacetonylpyridinium bromide (BPB) to study the jasmonate biosynthetic pathway of rice under salt stress and the premature senescence mutant. Adamec *et al.* [55] developed a new method to improve the identification and relative quantitation of ganoderic acids and other triterpenes in Ganoderma mushroom extracts.

Metabolic Labeling Method

Metabolic labeling incorporate detection or affinity tags into biomolecules in living cells. There are many advantages of metabolic labeling, such as low background, cell survival correlation, simple for cell labeling, few toxic and non-toxic byproducts produced [56]. As the earliest isotope labeling technology introduced into proteomics research, metabolic labeling includes $^{15}N/^{14}N$ labeling technology and the stable isotope labeling with amino acids in cell culture (SILAC) technology. In addition, there are reports on *in vivo* labeling using isotopes such as ^{13}C [57] and ^{34}S [58].

$^{15}N/^{14}N$ Labeling Technology

The first application of $^{15}N/^{14}N$ labeling was carried out by Oda *et al.* [59] on yeast cells, *Saccharomyces cerevisiae* cells, in cultures rich in ^{14}N or ^{15}N isotopes, respectively. The cells in two media were mixed in equal amounts, and proteins were extracted and digested. The expressed proteins under different growth conditions were determined by comparing the peak intensity ratios of the characteristic isotope peak pairs (^{14}N and ^{15}N labeled peptides) in MS/MS. 42 high-abundance proteins in *S. cerevisiae* cells were quantified with significant differences in the expression of G1 cyclinCLN2. Subsequently, the $^{15}N/^{14}N$ labeling technology was gradually applied to Nematode [60], Drosophila [60], and Arabidopsis [61].

However, the normal growth of cells can be affected to some extent by ^{15}N-rich media, and some biological systems are difficult to apply ^{15}N labeling in metabolism. It is difficult to obtain reliable data from relative quantitation in these circumstances. Incomplete ^{15}N labeling makes MS spectrum analysis more difficult [62].

SILAC Technology

In order to conquer disadvantages of $^{15}N/^{14}N$ labeling technology, Ong and his team [63] directly added stable isotope labeled amino acids to the cell culture system, which is called SILAC technology (Fig. **11**). During cell metabolism in culture, proteins were assembled with stable isotope labeled amino acids. The expressed proteins with stable isotope labeled amino acids were separated, digested, and quantitatively detected with MS/MS. SILAC greatly simplifies the processing complexity before MS analysis, and enhances data accuracy of relative quantitation [61a].

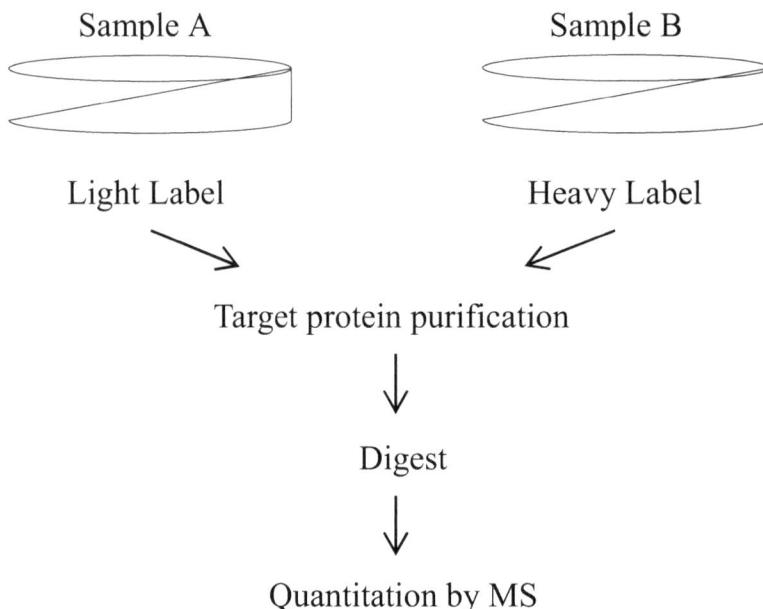

Fig. (11). General process of SILAC technology.

^{2}H, ^{13}C, and ^{15}N are stable isotopes commonly applied in SILAC technology. There will be an isotopic effect on the target component of ^{2}H labeling, resulting in a slight change in retention time during chromatographic separation and affecting the accuracy of the final analysis results [64].

At present, SILAC technology is more applied to cell systems and single-cell biological systems. For animal samples with long metabolic cycles and complex structures, the application of SILAC technology is greatly limited. When the plant cells are labeled by SILAC technology, incomplete labeling occurs because plant cells themselves can synthesize essential amino acids. Therefore, SILAC technology is not suitable for plant proteomics research.

Enzymatic Labeling Method

In 2001, Fenselau *et al*. [65] proposed ^{18}O labeling method. Two ^{18}O atoms are generally incorporated into the carboxyl termini of all tryptic peptides during the cleavage of all proteins in $H_2^{18}O$. Proteins in $H_2^{16}O$ are cleaved analogously with the carboxyl termini of the resulting peptides containing two ^{16}O atoms. Two groups of samples are treated to be analyzed by MS. The masses and isotope ratios of [18]O labeled and unlabeled peptides (differing by 4 Da) are used to identify and quantify proteins.

The mechanism of the +4 Da mass shift for the peptide hydrolysates by trypsin, Lys-C and Glu-C is that peptide hydrolysates continue to interact with proteases and undergo repeated binding/hydrolysis cycles, resulting in complete equilibration of two oxygen in carboxyl termini of hydrolysates with [18]O from $H_2{}^{18}O$ after proteolytic cleavage. In contrast, some proteases, such as chymotrypsin and Asp-N, do not accept hydrolyzed peptides as substrates after the cleavage step, resulting in single [18]O atom into the C-termini of hydrolyzed peptides with +2 Da mass shift [66].

Application of [18]O Labeling Method in Food Protein Research

Sha and his partners [67] used trypsin-catalyzed [18]O labeling method with HPLC-LTQ/Orbitrap MS/MS to study the characteristics of peptides in donkey-hide gelatin. Moreover, the method was applied to quantify donkey-hide gelatin and bovine-hide gelatin in gelatin mixtures.

The donkey-hide gelatin solution was centrifuged, filtered, and lyophilized for later use. Donkey-hide gelatin, bovine-hide gelatin and their mixtures (donkey-hide gelatin and bovine-hide gelatin were mixed in a mass ratio of 10:1, 1:1 and 1:10) were sequentially digested by Lys-C and trypsin, labeled with [18]O and analyzed by HPLC-LTQ/Orbitrap MS/MS. When Mascot score was more than 40, the peptides were identified as characteristic peptides of donkey-hide gelatin and bovine-hide gelatin [68].

Trypsin and Lys-C incorporated two [18]O atoms with +4 Da mass shift for the peptide hydrolysates, was further confirmed in this experiment. When the target gelatin content is less than 10%, 5 characteristic peptides of donkey-hide gelatin and 10 characteristic peptides of bovine-hide gelatin were identified and quantified, respectively, in the mixture of donkey-hide gelatin and bovine-hide gelatin (Table **2**). When the mass ratio of donkey-hide gelatin and bovine-hide gelatin was 1:10, 1:1 and 10:1, the ratio of donkey-hide gelatin and bovine-hide gelatin detected was 0.10, 1.02 and 9.72, respectively, which was close to the real values. Therefore, a precise quality control method of donkey-hide gelatin was setup with trypsin-catalyzed [18]O labeling combined with LC-MS/MS.

Table 2. Sequence of characteristic peptides of donkey-hide gelatin and bovine-hide gelatin.

Sequence of Characteristic Peptides of Donkey-Hide Gelatin	Sequence of Characteristic Peptides of Bovine-Hide Gelatin
α_1: [472] GE*PGPTGL*PGP*PGER [486]	α_1: [934] GAPGADGPAGA*PGTPGP*QGIAG*QR [957]
α_1: [1066] GEAGPAGPAGPIGPVGAR [1083]	α_1: [934]GA*PGADGPAGA*PGTPGP*QGIAG*QR [957]
α_1: [1062] SGDRGEAGPAGPAGPIGPVGAR [1083]	α_1:[1066] GETGPAGPAGPIGPVGAR [1083]
α_2: [881]GL*PGVAGSLGE*PGPLGIAGP*PGAR [904]	α_2: [326] GI*PGPVGAAGATGAR [340]
α_2: [977] GE*PGPVGSVGPVGAVGPR [994]	α_2: [620] GE*PGVVGA*PGTAGPSGPSGLPGER [643]
	α_2: [620] GE*PGVVGA*PGTAGPSGPSGL*PGER [643]
	α_2: [829] SGETGASGP*PGFVGEK [844]
	α_2: [977] GE*PGPAGAVGPAGAVGPR [994]
	α_2: [1066] IG*QPGAVGPAGIR [1078]
	α_2: [1066] IG*Q*PGAVGPAGIR [1078]

*P. represents hydroxylated proline and *Q. represents deacidification of glutamine.

^{18}O labeling method has a number of advantages over metabolic or chemical labeling [69]: (1) the reaction is simple and no additional processing steps are required to remove excess reagents, (2) H_2^{18}O is readily available and the volumes needed are economically justified, (3) every tryptic peptide (except for those proteins whose C-termini does not end with lysine or arginine) is labeled with the same mass label, (4) the mass difference provides sufficient separation between ^{18}O labeled peptides and ^{16}O labeled peptides, while maintaining identical ionization behavior as well as highly similar chromatographic properties, and (5) the method can label all tryptic peptides from any type of samples, such as tissues, urine samples, serum, *etc.*

^{18}O labeling method has several limitations in its application, including potentially incomplete labeling, back-exchange from ^{18}O to ^{16}O, and sample losses during labeling [70]. The most significant limitation is that trypsin catalyzes the back-exchange of ^{18}O with ^{16}O after labeling, leading to inaccurate results [70]. It is necessary to block trypsin hydrolytic activity. The common methods to inactivate trypsin are adjusting pH to 2 [69] and trypsin alkylating.

Introduction of Multiple Reaction Monitoring (MRM)

The use of isotope-labeled synthetic standards in quantitative MS has a long history [71]. It is now more widely applied as a method commonly referred to as AQUA (absolute quantitation of protein expression and PTM levels) [72]. A precise amount of a stable isotope-labeled standard peptide is added to the protein digestion, and the mass spectrometric signal of the labeled peptide is compared with that of the endogenous peptides in the sample. Since the amount of internal standard is known, and the ratio between amounts of internal standard and analytes can be determined by MS, the amount of the analyte peptides can be calculated. Through this procedure, absolute quantitation can be achieved. However, the AQUA method has many limitations. For example, it is unclear that how much is the appropriate amount of the labeled standard to be added, and which standard peptide is the best sequence to be added. Multiple reaction monitoring (MRM) can greatly improve the AQUA condition [73].

MRM, also known as selective reaction monitoring (SRM), applies triple quadrupole (QqQ) mass spectrometers to monitor both the mass of intact peptide and the mass of one or more specific fragment ions of that peptide during an LC-MS experiment [74]. The first mass analyzer (Q1) of these tandem instruments can selectively deliver ions of a specific m/z, corresponding to precursor ions of the target analyte. The precursor ions are then induced to fragments in a collision cell (q2) by collision-induced dissociation (CID) with neutral gas at appropriate pressure. This process produces fragment ions of the target analyte. The fragment ions with high ion intensity and specificity are again selectively delivered through a mass analyzer (Q3) (Fig. **12**). The transmitted ions finally reach the ion detector in mass spectrometers and the detected signals are recorded as an ion chromatogram for the precursor-fragment ion pair. The corresponding m/z pairs are called gas-phase transitions in MRM-MS experiments, which are predetermined during the experiment design stage [75]. MRM-MS can offer highly sensitive, specific, and cost-effective analysis for simultaneous quantitation of hundreds to several thousands of targeted peptides in a single experiment.

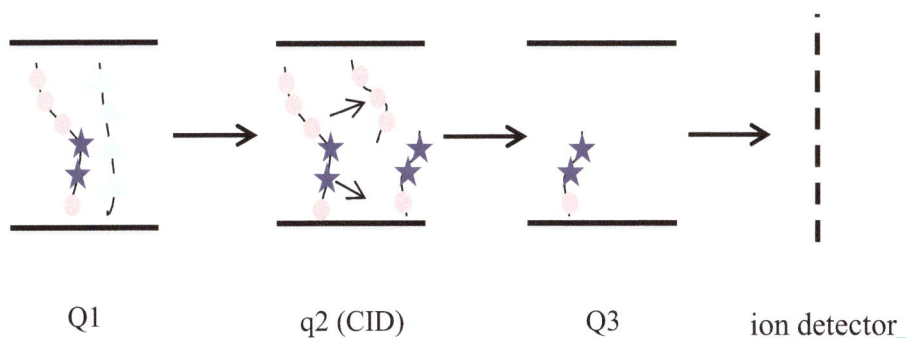

Q1   q2 (CID)   Q3   ion detector

Fig. (12). Schematic diagram of the multiple reaction monitoring (MRM) scanning technique on a triple quadrupole mass spectrometer.

Application of MRM in Proteins and Peptides Research

MRM method was first developed and validated with a standard protein (myoglobin). Subsequent experiments were performed for the absolute quantification of proteins and their modification states [72]. In experiments, internal standard peptides were synthesized with stable isotopes to mimic native peptides digested from proteins. These synthetic peptides could also be prepared with covalent modifications (*e.g.*, phosphorylation, methylation, and acetylation, *etc.*) that were chemically identical to naturally occurring PTMs. Such AQUA internal standard peptides were used to precisely quantify the absolute levels of proteins and post-translationally modified proteins by using MRM analysis in MS/MS. Then, peptide samples were digested by enzymes such as trypsin. Analyte and internal standard were analyzed by on-line microcapillary LC–MS/MS. Data were processed by integrating the appropriate peaks in an extracted ion chromatogram (EIC) for the native and internal standard, followed by calculation of the ratio of peak areas multiplied by the absolute amount of internal standard. At last, myoglobin in a yeast background and yeast silent information regulatory Sir2 and Sir4 proteins were analyzed. Through the same strategy, the cell cycle-dependent phosphorylation of Ser-1126 of human separase and human separase Ser-1501 *in vitro* kinase were also quantitatively determined.

To demonstrate the AQUA method, a measurement of the lysine 48 (K^{48}) branched chain of polyubiquitin was presented [76]. The PTM of a protein by a polyubiquitin chain linked through K^{48} was recognized by the 26S proteasome as a signal for protein degradation [77]. Digestion with trypsin to produce a branched peptide of reasonable size for analysis was essential, since polyubiquitin chains were large PTMs. In the case of K^{48}-linked chains, this digestion produced a peptide with sequence LIFAGK^{48} (GG) Q*L* EDGR, where K^{48} was modified with a di-glycine residue attached through an isopeptide bond to the epsilon-amino group of lysine. An internal standard peptide with this sequence was synthesized and enriched in stable isotopes at leucine 50. The amount of this branched peptide formed by trypsin digestion directly correlated with the amount of K^{48} polyubiquitin branched chains attached to proteins. Using this ubiquitin-AQUA method, it is possible to quantify polyubiquitin chains of various conformations [76].

The capabilities of high-resolution MS instruments for circumventing the complex sample preparation, which were needed for sensitive LC-MS/MS-based quantification of monoclonal antibodies (mAbs), were investigated [78]. The ability of three different LC-MS platforms for absolute quantification of a representative mAb Rendomab-B1 in serum and plasma was compared. MS/MS acquisitions were developed and systematically examined for 1) a low-resolution

QqQ operated in selected reaction monitoring (SRM) acquisition mode, 2) a high-resolution hybrid Quadrupole-Orbitrap (Q-Orbitrap) operated in parallel reaction monitoring (PRM) acquisition mode and 3) a high-resolution hybrid Quadrupole Time-of-flight (Q-TOF) operated in SRM acquisition mode with enhanced duty cycle function. The result showed that the sensitivity of the high-resolution Q-Orbitrap and Q-TOF methods was significantly higher (LOD of 80 ng/mL) in serum/plasma samples than the low-resolution QqQ method [78]. Moreover, compared to traditional ligand binding assay (LBA), the developed high-resolution MS method will have a better practical application in determining the pharmacokinetic (PK) properties of therapeutic mAbs.

A method for the quantitation of P-gp and Na^+/K^+ ATPase α-subunit isoforms (Atp1a 1, 2 and 3) at the blood-brain barrier (BBB) by UHPLC–MS/MS (MRM) using the AQUA strategy with several peptides per protein was developed by Gomez-Zepeda and his team [79]. To improve the confidence on quantitation, all the assays had been comprehensively validated in terms of linearity, accuracy, precision, digestion efficiency and peptide stability. In their work, this method was successfully applied to the determination of proteins in rat mouse kidney, mouse cortical vessels and rat cortical micro vessels. What's more, the different levels obtained for each peptide highlighted the importance and difficulty of selecting a peptide for protein quantitation.

Label-Free Quantitation Method

Label-free quantitation method, which does not use a stable isotope that chemically binds and labels the protein [80], is cost-effective and only requires minimal sample preparation. In contrast to labeled quantitation methods, samples must be measured separately in a label-free experiment. The extracted peptide signals are then identified across LC-MS measurements using their coordinates on the m/z ratio and retention time dimensions. There are two commonly used quantitative schemes in label-free quantitative method: 1. approach of mass spectral peak intensities, which is based on good correlation between proteins/peptides amount and the peak intensities, and 2. spectral counting approach, which is based on spectral counts of peptides unique to a specific protein [81].

Application of Label-Free Quantitation Method in Food Protein Research

The effect of melatonin on promoting fruit ripening and anthocyanin accumulation was studied upon postharvest in tomato, using a label-free quantitation method [82].

The water-treated tomatoes were setup as control check (CK) group and 50 μM melatonin-treated tomatoes were setup as M50 group. Proteins were extracted from the tomatoes fruits 5 days after the melatonin treatment (DAT) [83] and were digested with trypsin. Tryptic digestions were desalted with C18 Cartridges and analyzed by LC-MS/MS. The MS data were analyzed with MaxQuant software. The MS data were searched against the Uniprot solanum lycopersicum database in UniProtKB.

3070 and 3087 extracted proteins from CK and M50 groups were identified, respectively. There were 2898 proteins identified in both tomato groups. Among the identified proteins, 241 differentially expressed proteins were quantified, 136 were up-regulated and 105 were down-regulated, indicating that melatonin influenced tomato ripening at the protein level. For example, a senescence-related protein was downregulated in the M50 fruit, while a cell apoptosis inhibitor (API5) protein was upregulated. In addition, peroxidases (POD9, POD12, peroxidase p7-like) and catalase (CAT3) significantly increased in the M50 fruits. These proteins were involved in several ripening-related pathways, including cell wall metabolism, oxidative phosphorylation, carbohydrate metabolism, and fatty acid metabolism. Moreover, the application of exogenous melatonin increased 8 proteins associated with anthocyanin accumulation during fruit ripening. Analysis of anthocyanin content from M50 showed that its level significantly increased by 52%, 48%, and 50% at 5, 8, and 13 DAT, respectively. At the protein concentration level, melatonin-mediated fruit ripening as well as the anthocyanin formation process in tomato fruit were possibly revealed in physiological and molecular mechanisms during fruit ripening.

Wang *et al*. [84] studied the regulatory mechanisms of Se response in rice grains by carrying out a comparative proteomics study in Se-enriched rice grains using two-dimensional gel electrophoresis (2-DE)-coupled MALDI-TOF/TOF MS and 1-DE/LC−FTICR−MS-coupled label-free quantitation. Comparing Se treatment and control, 62 and 250 differentially expressed proteins were identified from 2-DE and 1-DE, respectively. These proteins, which involved in metabolism, cell redox regulation, and seed nutritional storage, were the most highly affected ones by Se accumulation. Late embryogenesis abundant proteins as well as proteins involved in sucrose synthesis and other metabolism pathways were up-regulated, which may lead to the earlier maturation and higher yield of Se-enriched rice. In addition, six proteins containing seleno-amino acid modification were identified in rice at first time. Novel insights into Se response in rice grains at the proteome level were provided, which were useful for the production of Se-enriched rice in the future.

Comparison of Two Approaches of Label-Free Quantitation Method

Spectral counting approach simply counts the number of tandem spectra from the same protein identified in LC-MS/MS experiments. With more complicated techniques, the peak intensities approach is generally more reliable than the spectral counting approach because m/z, peak area, and retention time for each ion in all experiments need to be well aligned (Table **3**). Strict calibration of chromatograph and mass spectrometer has to be guaranteed for high reproducibility between LC-MS/MS runs, which is a key factor for reliable quantitation [81].

Table 3. Comparison of peak intensities approach and spectral counting approach.

	Peak Intensities Approach	**Spectral Counting Approach**
Accuracy [85]	high	low
Sensitivity [85, 86]	low	high
Linearity [86]	low	high
Dynamic range [87]	narrow	wide
Repeatability [87]	poor	good
Data processing complexity [88]	hard	easy

Comparison Between Labeled Quantitation and Label-Free Quantitation Methods

Label-free quantitation method can directly analyze large-scale samples of proteins and peptides with independent injections [89]. Label-free quantitation method has several advantages over labeled quantitation methods: it is cost-effective without expensive labeling reagents; it is time-saving without tedious labeling steps [90]. Label-free quantitation method is insensitive to technical error and highly sensitive to identify thousands of proteins from complex samples such as bodily fluids (blood, plasma, saliva, and urine), cell lines, and tissues in MS analysis [91]. The characteristics and applications of quantitative MS methods are listed in Table **4** [73].

Table 4. Characteristics and applications of quantitative mass spectrometry methods.

Method		Application	Accuracy (Process)	Quantitative Proteome Coverage	Linear Dynamic Range[a]
Chemical labeling	ICAT	Medium to complex biochemical workflows Comparison of 2–3 samples	+++	++	1-2 logs
	iTRAQ	Medium complexity biochemical workflows Comparison of 2–8 samples	+++	++	1-2 logs
	Carboxyl labeling method	Medium complexity biochemical workflows Comparison of 2–3 samples	++	++	2 logs
Metabolic labeling	SILAC	Complex biochemical workflows Comparison of 2–3 samples Cell culture systems only	++	++	2 logs
Enzymatic labeling	18 O labeling method	Medium complexity biochemical Workflows Comparison of 2 samples	++	++	1-2 logs
Label-free quantitation	peak intensities approach	Simple biochemical workflows Whole proteome analysis Comparison of multiple samples	+	+++	2-3 logs
	spectral counting approach	Simple biochemical workflows Whole proteome analysis Comparison of multiple samples	+	+++	2-3 logs

[a] In MRM mode, dynamic range may be extended to 4–5 logs [92].

OUTLOOK

Qualitative and quantitative analysis of proteins and peptides has rapidly developed in food science, nutrition science, and life science. Numerous methods have emerged over the years for the analysis of various proteomes. As we have discussed in this Chapter, MS-based qualitative and quantitative analysis has particular advantages. Moreover, it is necessary to develop various bioinformatics and statistical tools for simplifying the complexity of data as well as managing huge data generated.

CONSENT FOR PUBLICATION

Not applicable.

CONFLICT OF INTEREST

The authors confirm that this chapter contents have no conflict of interest.

ACKNOWLEDGEMENTS

Declared none.

REFERENCES

[1]   Tanaka, K.; Waki, H.; Ido, Y.; Akita, S.; Yoshida, Y.; Yoshida, T.; Matsuo, T. Protein and polymer analyses up to m/z 100 000 by laser ionization time-of-flight mass spectrometry. *Rapid Commun. Mass Spectrom.,* **1988**, *2*(8), 151-153.
      [http://dx.doi.org/10.1002/rcm.1290020802]

[2]   Fenn, J.B.; Mann, M.; Meng, C.K.; Wong, S.F.; Whitehouse, C.M. Electrospray ionization for mass spectrometry of large biomolecules. *Science,* **1989**, *246*(4926), 64-71.
      [http://dx.doi.org/10.1126/science.2675315] [PMID: 2675315]

[3]   Pramanik, B.N.; Ganguly, A.K.; Gross, M.L. Applied Electrospray Mass Spectrometry. Practical Spectroscopy Series. *J. Am. Chem. Soc.,* **2002**, *124*(45), 13638-13639.

[4]   a) Eng, J.K.; McCormack, A.L.; Yates, J.R. An approach to correlate tandem mass spectral data of peptides with amino acid sequences in a protein database. *J. Am. Soc. Mass Spectrom.,* **1994**, *5*(11), 976-989.
      [http://dx.doi.org/10.1016/1044-0305(94)80016-2] [PMID: 24226387] b) Perkins, D.N.; Pappin, D.J.C.; Creasy, D.M.; Cottrell, J.S. Probability-based protein identification by searching sequence databases using mass spectrometry data. *Electrophoresis,* **1999**, *20*(18), 3551-3567.
      [http://dx.doi.org/10.1002/(SICI)1522-2683(19991201)20:18<3551::AID-ELPS3551>3.0.CO;2-2]
      [PMID: 10612281] c) Clauser, K.R.; Baker, P.; Burlingame, A.L. Role of accurate mass measurement (+/- 10 ppm) in protein identification strategies employing MS or MS/MS and database searching. *Anal. Chem.,* **1999**, *71*(14), 2871-2882.
      [http://dx.doi.org/10.1021/ac9810516] [PMID: 10424174] d) Field, H.I.; Fenyö, D.; Beavis, R.C. RADARS, a bioinformatics solution that automates proteome mass spectral analysis, optimises protein identification, and archives data in a relational database. *Proteomics,* **2002**, *2*(1), 36-47.
      [http://dx.doi.org/10.1002/1615-9861(200201)2:1<36::AID-PROT36>3.0.CO;2-W]   [PMID: 11788990] e) Craig, R.; Beavis, R.C. TANDEM: matching proteins with tandem mass spectra. *Bioinformatics,* **2004**, *20*(9), 1466-1467.
      [http://dx.doi.org/10.1093/bioinformatics/bth092] [PMID: 14976030] f) Geer, L.Y.; Markey, S.P.; Kowalak, J.A.; Wagner, L.; Xu, M.; Maynard, D.M.; Yang, X.; Shi, W.; Bryant, S.H. Open mass spectrometry search algorithm. *J. Proteome Res.,* **2004**, *3*(5), 958-964.
      [http://dx.doi.org/10.1021/pr0499491] [PMID: 15473683] g) Sadygov, R.G.; Yates, J.R., III A hypergeometric probability model for protein identification and validation using tandem mass spectral data and protein sequence databases. *Anal. Chem.,* **2003**, *75*(15), 3792-3798.
      [http://dx.doi.org/10.1021/ac034157w] [PMID: 14572045]

[5]   Zhang, N.; Aebersold, R.; Schwikowski, B.; Prob, I.D. ProbID: a probabilistic algorithm to identify peptides through sequence database searching using tandem mass spectral data. *Proteomics,* **2002**, *2*(10), 1406-1412.
      [http://dx.doi.org/10.1002/1615-9861(200210)2:10<1406::AID-PROT1406>3.0.CO;2-9]   [PMID: 12422357]

[6]   Allet, N.; Barrillat, N.; Baussant, T.; Boiteau, C.; Botti, P.; Bougueleret, L.; Budin, N.; Canet, D.; Carraud, S.; Chiappe, D.; Christmann, N.; Colinge, J.; Cusin, I.; Dafflon, N.; Depresle, B.; Fasso, I.; Frauchiger, P.; Gaertner, H.; Gleizes, A.; Gonzalez-Couto, E.; Jeandenans, C.; Karmime, A.; Kowall, T.; Lagache, S.; Mahé, E.; Masselot, A.; Mattou, H.; Moniatte, M.; Niknejad, A.; Paolini, M.; Perret, F.; Pinaud, N.; Ranno, F.; Raimondi, S.; Reffas, S.; Regamey, P.O.; Rey, P.A.; Rodriguez-Tomé, P.;

Rose, K.; Rossellat, G.; Saudrais, C.; Schmidt, C.; Villain, M.; Zwahlen, C. In vitro and in silico processes to identify differentially expressed proteins. *Proteomics,* **2004**, *4*(8), 2333-2351.
[http://dx.doi.org/10.1002/pmic.200300840] [PMID: 15274127]

[7]   Kou, X.H.; Gao, J.; Xue, Z.H.; Zhang, Z.J. Purification and identification of antioxidant peptides from chickpea (Cicer arietinum L.) albumin hydrolysates. *Lebensm. Wiss. Technol.,* **2013**, *50*(2), 591-598.
[http://dx.doi.org/10.1016/j.lwt.2012.08.002]

[8]   Xue, Z.; Yu, W.; Liu, Z.; Wu, M.; Kou, X.; Wang, J. Preparation and antioxidative properties of a rapeseed (Brassica napus) protein hydrolysate and three peptide fractions. *J. Agric. Food Chem.,* **2009**, *57*(12), 5287-5293.
[http://dx.doi.org/10.1021/jf900860v] [PMID: 19432452]

[9]   Xue, Z.; Liu, Z.; Wu, M.; Zhuang, S.; Yu, W. Effect of rapeseed peptide on DNA damage and apoptosis in Hela cells. *Exp. Toxicol. Pathol.,* **2010**, *62*(5), 519-523.
[http://dx.doi.org/10.1016/j.etp.2009.06.013] [PMID: 19640692]

[10]   Nanjo, F.; Goto, K.; Seto, R.; Suzuki, M.; Sakai, M.; Hara, Y. Scavenging effects of tea catechins and their derivatives on 1,1-diphenyl-2-picrylhydrazyl radical. *Free Radic. Biol. Med.,* **1996**, *21*(6), 895-902.
[http://dx.doi.org/10.1016/0891-5849(96)00237-7] [PMID: 8902534]

[11]   Halliwell, B.; Gutteridge, J.M.C.; Aruoma, O.I. The deoxyribose method: a simple "test-tube" assay for determination of rate constants for reactions of hydroxyl radicals. *Anal. Biochem.,* **1987**, *165*(1), 215-219.
[http://dx.doi.org/10.1016/0003-2697(87)90222-3] [PMID: 3120621]

[12]   Re, R.; Pellegrini, N.; Proteggente, A.; Pannala, A.; Yang, M.; Rice-Evans, C. RE. Antioxidant activity applying an improved ABTS radical cation decolorization assay. *Free Radic. Biol. Med.,* **1999**, *26*(9-10), 1231-1237.
[http://dx.doi.org/10.1016/S0891-5849(98)00315-3] [PMID: 10381194]

[13]   Benzie, I.F.F.; Strain, J.J. The Ferric Reducing Ability of Plasma (FRAP) as a Measure of. The ferric reducing ability of plasma (FRAP) as a measure of "antioxidant power": the FRAP assay. *Anal. Biochem.,* **1996**, *239*(1), 70-76.
[http://dx.doi.org/10.1006/abio.1996.0292] [PMID: 8660627]

[14]   Li, Y.H.; Jiang, B.; Zhang, T.; Mu, W.M.; Liu, J. Antioxidant and free radical-scavenging activities of chickpea protein hydrolysate (CPH). *Food Chem.,* **2008**, *106*(2), 444-450.
[http://dx.doi.org/10.1016/j.foodchem.2007.04.067]

[15]   Zhang, J.H.; Zhang, H.; Wang, L.; Guo, X.N. Isolation and identification of antioxidative peptides from rice endosperm protein enzymatic hydrolysate by consecutive chromatography and MALDI-TOF/TOF MS/MS. *Food Chem.,* **2009**, *119*(1), 226-234.
[http://dx.doi.org/10.1016/j.foodchem.2009.06.015]

[16]   Wang, Y.; Chen, H.; Wang, X.; Li, S.; Chen, Z.; Wang, J.; Liu, W. Isolation and identification of a novel peptide from zein with antioxidant and antihypertensive activities. *Food Funct.,* **2015**, *6*(12), 3799-3806.
[http://dx.doi.org/10.1039/C5FO00815H] [PMID: 26422315]

[17]   Nagao, J.; Miyashita, M.; Nakagawa, Y.; Miyagawa, H. Chemical synthesis of La1 isolated from the venom of the scorpion Liocheles australasiae and determination of its disulfide bonding pattern. *J. Pept. Sci.,* **2015**, *21*(8), 636-643.
[http://dx.doi.org/10.1002/psc.2778] [PMID: 25919411]

[18]   Willems, P.; Horne, A.; Van Parys, T.; Goormachtig, S.; De Smet, I.; Botzki, A.; Van Breusegem, F.; Gevaert, K. The Plant PTM Viewer, a central resource for exploring plant protein modifications. *Plant J.,* **2019**, *99*(4), 752-762.
[http://dx.doi.org/10.1111/tpj.14345] [PMID: 31004550]

[19]   Lam, M.P.Y.; Scruggs, S.B.; Kim, T-Y.; Zong, C.; Lau, E.; Wang, D.; Ryan, C.M.; Faull, K.F.; Ping,

P. An MRM-based workflow for quantifying cardiac mitochondrial protein phosphorylation in murine and human tissue. *J. Proteomics,* **2012**, *75*(15), 4602-4609.
[http://dx.doi.org/10.1016/j.jprot.2012.02.014] [PMID: 22387130]

[20]   Vermassen, T.; Speeckaert, M.M.; Lumen, N.; Rottey, S.; Delanghe, J.R. Glycosylation of prostate specific antigen and its potential diagnostic applications. *Clin. Chim. Acta,* **2012**, *413*(19-20), 1500-1505.
[http://dx.doi.org/10.1016/j.cca.2012.06.007] [PMID: 22722018]

[21]   He, Y.; Korboukh, I.; Jin, J.; Huang, J. Targeting protein lysine methylation and demethylation in cancers. *Acta Biochim. Biophys. Sin. (Shanghai),* **2012**, *44*(1), 70-79.
[http://dx.doi.org/10.1093/abbs/gmr109] [PMID: 22194015]

[22]   Sustáčková, G.; Kozubek, S.; Stixová, L.; Legartová, S.; Matula, P.; Orlova, D.; Bártová, E. Acetylation-dependent nuclear arrangement and recruitment of BMI1 protein to UV-damaged chromatin. *J. Cell. Physiol.,* **2012**, *227*(5), 1838-1850.
[http://dx.doi.org/10.1002/jcp.22912] [PMID: 21732356]

[23]   Zhu, Y.S.; FitzGerald, R.J. Direct nanoHPLC-ESI-QTOF MS/MS analysis of tryptic caseinophosphopeptides. *Food Chem.,* **2010**, *123*(3), 753-759.
[http://dx.doi.org/10.1016/j.foodchem.2010.05.046]

[24]   Adamson, N.J.; Reynolds, E.C. Characterization of tryptic casein phosphopeptides prepared under industrially relevant conditions. *Biotechnol. Bioeng.,* **1995**, *45*(3), 196-204.
[http://dx.doi.org/10.1002/bit.260450303] [PMID: 18623138]

[25]   Zhu, Y.S.; FitzGerald, R.J. Caseinophosphopeptide enrichment and identification. *Int. J. Food Sci. Technol.,* **2012**, *47*(10), 2235-2242.
[http://dx.doi.org/10.1111/j.1365-2621.2012.03094.x]

[26]   Hua, S.; Nwosu, C.C.; Strum, J.S.; Seipert, R.R.; An, H.J.; Zivkovic, A.M.; German, J.B.; Lebrilla, C.B. Site-specific protein glycosylation analysis with glycan isomer differentiation. *Anal. Bioanal. Chem.,* **2012**, *403*(5), 1291-1302.
[http://dx.doi.org/10.1007/s00216-011-5109-x] [PMID: 21647803]

[27]   Wu, C.C.; MacCoss, M.J.; Howell, K.E.; Matthews, D.E.; Yates, J.R., III Metabolic labeling of mammalian organisms with stable isotopes for quantitative proteomic analysis. *Anal. Chem.,* **2004**, *76*(17), 4951-4959.
[http://dx.doi.org/10.1021/ac049208j] [PMID: 15373428]

[28]   Ong, S.E.; Mann, M. Mass spectrometry-based proteomics turns quantitative. *Nat. Chem. Biol.,* **2005**, *1*(5), 252-262.
[http://dx.doi.org/10.1038/nchembio736] [PMID: 16408053]

[29]   Gygi, S.P.; Rist, B.; Gerber, S.A.; Turecek, F.; Gelb, M.H.; Aebersold, R. Quantitative analysis of complex protein mixtures using isotope-coded affinity tags. *Nat. Biotechnol.,* **1999**, *17*(10), 994-999.
[http://dx.doi.org/10.1038/13690] [PMID: 10504701]

[30]   Zhang, R.; Sioma, C.S.; Wang, S.; Regnier, F.E. Fractionation of isotopically labeled peptides in quantitative proteomics. *Anal. Chem.,* **2001**, *73*(21), 5142-5149.
[http://dx.doi.org/10.1021/ac010583a] [PMID: 11721911]

[31]   Kim, J.R.; Yoon, H.W.; Kwon, K.S.; Lee, S.R.; Rhee, S.G. Identification of proteins containing cysteine residues that are sensitive to oxidation by hydrogen peroxide at neutral pH. *Anal. Biochem.,* **2000**, *283*(2), 214-221.
[http://dx.doi.org/10.1006/abio.2000.4623] [PMID: 10906242]

[32]   Sethuraman, M.; McComb, M.E.; Huang, H.; Huang, S.; Heibeck, T.; Costello, C.E.; Cohen, R.A. Isotope-coded affinity tag (ICAT) approach to redox proteomics: identification and quantitation of oxidant-sensitive cysteine thiols in complex protein mixtures. *J. Proteome Res.,* **2004**, *3*(6), 1228-1233.
[http://dx.doi.org/10.1021/pr049887e] [PMID: 15595732]

[33]   Islam, N.; Tsujimoto, H.; Hirano, H. Wheat proteomics: relationship between fine chromosome deletion and protein expression. *Proteomics,* **2003**, *3*(3), 307-316.
[http://dx.doi.org/10.1002/pmic.200390044] [PMID: 12627384]

[34]   Flory, M.R.; Lee, H.; Bonneau, R.; Mallick, P.; Serikawa, K.; Morris, D.R.; Aebersold, R. Quantitative proteomic analysis of the budding yeast cell cycle using acid-cleavable isotope-coded affinity tag reagents. *Proteomics,* **2006**, *6*(23), 6146-6157.
[http://dx.doi.org/10.1002/pmic.200600159] [PMID: 17133367]

[35]   Keller, A.; Nesvizhskii, A.I.; Kolker, E.; Aebersold, R. Empirical statistical model to estimate the accuracy of peptide identifications made by MS/MS and database search. *Anal. Chem.,* **2002**, *74*(20), 5383-5392.
[http://dx.doi.org/10.1021/ac025747h] [PMID: 12403597]

[36]   Pierce, A.; Unwin, R.D.; Evans, C.A.; Griffiths, S.; Carney, L.; Zhang, L.; Jaworska, E.; Lee, C.F.; Blinco, D.; Okoniewski, M.J.; Miller, C.J.; Bitton, D.A.; Spooncer, E.; Whetton, A.D. Eight-channel iTRAQ enables comparison of the activity of six leukemogenic tyrosine kinases. *Mol. Cell. Proteomics,* **2008**, *7*(5), 853-863.
[http://dx.doi.org/10.1074/mcp.M700251-MCP200] [PMID: 17951628]

[37]   Ross, P.L.; Huang, Y.N.; Marchese, J.N.; Williamson, B.; Parker, K.; Hattan, S.; Khainovski, N.; Pillai, S.; Dey, S.; Daniels, S.; Purkayastha, S.; Juhasz, P.; Martin, S.; Bartlet-Jones, M.; He, F.; Jacobson, A.; Pappin, D.J. Multiplexed protein quantitation in Saccharomyces cerevisiae using amine-reactive isobaric tagging reagents. *Mol. Cell. Proteomics,* **2004**, *3*(12), 1154-1169.
[http://dx.doi.org/10.1074/mcp.M400129-MCP200] [PMID: 15385600]

[38]   Schwacke, J.H.; Hill, e.g.; Krug, E.L.; Comte-Walters, S.; Schey, K.L. iQuantitator: a tool for protein expression inference using iTRAQ. *BMC Bioinformatics,* **2009**, *10*(1), 342-357.
[http://dx.doi.org/10.1186/1471-2105-10-342] [PMID: 19835628]

[39]   Golovan, S.P.; Hakimov, H.A.; Verschoor, C.P.; Walters, S.; Gadish, M.; Elsik, C.; Schenkel, F.; Chiu, D.K.Y.; Forsberg, C.W. Analysis of Sus scrofa liver proteome and identification of proteins differentially expressed between genders, and conventional and genetically enhanced lines. *Comp. Biochem. Physiol. Part D Genomics Proteomics,* **2008**, *3*(3), 234-242.
[http://dx.doi.org/10.1016/j.cbd.2008.05.001] [PMID: 20483222]

[40]   Shilov, I.V.; Seymour, S.L.; Patel, A.A.; Loboda, A.; Tang, W.H.; Keating, S.P.; Hunter, C.L.; Nuwaysir, L.M.; Schaeffer, D.A. The Paragon Algorithm, a next generation search engine that uses sequence temperature values and feature probabilities to identify peptides from tandem mass spectra. *Mol. Cell. Proteomics,* **2007**, *6*(9), 1638-1655.
[http://dx.doi.org/10.1074/mcp.T600050-MCP200] [PMID: 17533153]

[41]   Liu, M.X.; Zhao, Y.; Zhang, M. Functional Proteomics Study of northeast Chinese sika deer antlers. *Shi Zhen National Medicine National Medicine,* **2018**, *29*(2), 322-324.

[42]   Zhu, Y.S.; Kalyankar, P.; FitzGerald, R.J. Relative quantitation analysis of the substrate specificity of glutamyl endopeptidase with bovine α-caseins. *Food Chem.,* **2015**, *167*, 463-467.
[http://dx.doi.org/10.1016/j.foodchem.2014.07.017] [PMID: 25149012]

[43]   Zhu, Y.S.; Kalyankar, P.; FitzGerald, R.J. Quantitative analysis of bovine β-casein hydrolysates obtained using glutamyl endopeptidase. *Lebensm. Wiss. Technol.,* **2015**, *63*(2), 1334-1338.
[http://dx.doi.org/10.1016/j.lwt.2015.04.021]

[44]   Zeng, R.; Farooq, M.U.; Wang, L.; Su, Y.; Zheng, T.; Ye, X.; Jia, X.; Zhu, J. Study on Differential Protein Expression in Natural Selenium-Enriched and Non-Selenium-Enriched Rice Based on iTRAQ Quantitative Proteomics. *Biomolecules,* **2019**, *9*(4), 130-147.
[http://dx.doi.org/10.3390/biom9040130] [PMID: 30935009]

[45]   Chen, L.; Huang, Y.; Xu, M.; Cheng, Z.; Zhang, D.; Zheng, J. iTRAQ-Based Quantitative Proteomics Analysis of Black Rice Grain Development Reveals Metabolic Pathways Associated with Anthocyanin Biosynthesis. *PLoS One,* **2016**, *11*(7)e0159238

[http://dx.doi.org/10.1371/journal.pone.0159238] [PMID: 27415428]

[46]   Yang, Q.S.; Wu, J.H.; Li, C.Y.; Wei, Y.R.; Sheng, O.; Hu, C.H.; Kuang, R.B.; Huang, Y.H.; Peng, X.X.; McCardle, J.A.; Chen, W.; Yang, Y.; Rose, J.K.; Zhang, S.; Yi, G.J. Quantitative proteomic analysis reveals that antioxidation mechanisms contribute to cold tolerance in plantain (Musa paradisiaca L.; ABB Group) seedlings. *Mol. Cell. Proteomics,* **2012**, *11*(12), 1853-1869.
       [http://dx.doi.org/10.1074/mcp.M112.022079] [PMID: 22982374]

[47]   Ming, T.Y.; Li, J.J.; Huo, P.; Wei, Y.L.; Chen, X.H. Analysis of Free Amino Acids in Russula griseocarnosa Harvested at Different Stages of Maturity Using iTRAQ-LC-MS/MS. *Food Anal. Methods,* **2014**, *7*(9), 1816-1823.
       [http://dx.doi.org/10.1007/s12161-014-9817-7]

[48]   Goodlett, D.R.; Keller, A.; Watts, J.D.; Newitt, R.; Yi, E.C.; Purvine, S.; Eng, J.K.; von Haller, P.; Aebersold, R.; Kolker, E. Differential stable isotope labeling of peptides for quantitation and de novo sequence derivation. *Rapid Commun. Mass Spectrom.,* **2001**, *15*(14), 1214-1221.
       [http://dx.doi.org/10.1002/rcm.362] [PMID: 11445905]

[49]   a) Annesley, T.M. Ion suppression in mass spectrometry. *Clin. Chem.,* **2003**, *49*(7), 1041-1044.
       [http://dx.doi.org/10.1373/49.7.1041] [PMID: 12816898] b) Constantopoulos, T.L.; Jackson, G.S.; Enke, C.G. Effects of salt concentration on analyte response using electrospray ionization mass spectrometry. *J. Am. Soc. Mass Spectrom.,* **1999**, *10*(7), 625-634.
       [http://dx.doi.org/10.1016/S1044-0305(99)00031-8] [PMID: 10384726] c) Sterner, J.L.; Johnston, M.V.; Nicol, G.R.; Ridge, D.P. Signal suppression in electrospray ionization Fourier transform mass spectrometry of multi-component samples. *J. Mass Spectrom.,* **2000**, *35*(3), 385-391.
       [http://dx.doi.org/10.1002/(SICI)1096-9888(200003)35:3<385::AID-JMS947>3.0.CO;2-O]   [PMID: 10767768] d) Tang, L.; Kebarle, P. Dependence of ion intensity in electrospray mass spectrometry on the concentration of the analytes in the electrosprayed solution. *Anal. Chem.,* **1993**, *65*(24), 3654-3668.
       [http://dx.doi.org/10.1021/ac00072a020]

[50]   Ou, Y.Y.; Sun, X.H.; Yan, J.; Zhu, J.F. Progress in relative quantitative analysis of biological molecules with stable isotope labeling. *Chin. Sci. Bull.,* **2013**, *58*(27), 2762-2778.
       [http://dx.doi.org/10.1360/972013-269]

[51]   Lamos, S.M.; Shortreed, M.R.; Frey, B.L.; Belshaw, P.J.; Smith, L.M. Relative quantification of carboxylic acid metabolites by liquid chromatography-mass spectrometry using isotopic variants of cholamine. *Anal. Chem.,* **2007**, *79*(14), 5143-5149.
       [http://dx.doi.org/10.1021/ac062416m] [PMID: 17563114]

[52]   Yang, W.C.; Adamec, J.; Regnier, F.E. Enhancement of the LC/MS analysis of fatty acids through derivatization and stable isotope coding. *Anal. Chem.,* **2007**, *79*(14), 5150-5157.
       [http://dx.doi.org/10.1021/ac070311t] [PMID: 17492837]

[53]   a) Yang, W.C.; Mirzaei, H.; Liu, X.; Regnier, F.E. Enhancement of amino acid detection and quantification by electrospray ionization mass spectrometry. *Anal. Chem.,* **2006**, *78*(13), 4702-4708.
       [http://dx.doi.org/10.1021/ac0600510] [PMID: 16808485] b) Johnson, D.W. Alkyldimethylaminoethyl ester iodides for improved analysis of fatty acids by electrospray ionization tandem mass spectrometry. *Rapid Commun. Mass Spectrom.,* **2000**, *14*(21), 2019-2024.
       [http://dx.doi.org/10.1002/1097-0231(20001115)14:21<2019::AID-RCM121>3.0.CO;2-2]   [PMID: 11085412] c) Barry, S.J.; Carr, R.M.; Lane, S.J.; Leavens, W.J.; Monté, S.; Waterhouse, I. Derivatisation for liquid chromatography/electrospray mass spectrometry: synthesis of pyridinium compounds and their amine and carboxylic acid derivatives. *Rapid Commun. Mass Spectrom.,* **2003**, *17*(6), 603-620.
       [http://dx.doi.org/10.1002/rcm.957] [PMID: 12621624] d) Mirzaei, H.; Regnier, F. Enhancing electrospray ionization efficiency of peptides by derivatization. *Anal. Chem.,* **2006**, *78*(12), 4175-4183.
       [http://dx.doi.org/10.1021/ac0602266] [PMID: 16771548]

[54]   Huang, Y.Q.; Liu, J.Q.; Gong, H.; Yang, J.; Li, Y.; Feng, Y-Q. Use of isotope mass probes for metabolic analysis of the jasmonate biosynthetic pathway. *Analyst (Lond.),* **2011**, *136*(7), 1515-1522.
       [http://dx.doi.org/10.1039/c0an00736f] [PMID: 21331428]

[55]   Adamec, J.; Jannasch, A.; Dudhgaonkar, S.; Jedinak, A.; Sedlak, M.; Sliva, D. Development of a new method for improved identification and relative quantification of unknown metabolites in complex samples: determination of a triterpenoid metabolic fingerprint for the in situ characterization of Ganoderma bioactive compounds. *J. Sep. Sci.,* **2009**, *32*(23-24), 4052-4058.
[http://dx.doi.org/10.1002/jssc.200900496] [PMID: 19937965]

[56]   Han, S.S.; Shim, H.E.; Park, S.J.; Kim, B-C.; Lee, D.E.; Chung, H.M.; Moon, S.H.; Kang, S.W. Safety and Optimization of Metabolic Labeling of Endothelial Progenitor Cells for Tracking. *Sci. Rep.,* **2018**, *8*(1), 13212.
[http://dx.doi.org/10.1038/s41598-018-31594-0] [PMID: 30181604]

[57]   Chen, W.P.; Yang, X.Y.; Harms, G.L.; Gray, W.M.; Hegeman, A.D.; Cohen, J.D. An automated growth enclosure for metabolic labeling of Arabidopsis thaliana with 13C-carbon dioxide - an *in vivo* labeling system for proteomics and metabolomics research. *Proteome Sci.,* **2011**, *9*(1), 9.
[http://dx.doi.org/10.1186/1477-5956-9-9] [PMID: 21310072]

[58]   Wu, Z.L.; Lech, M. Modification degrees at specific sites on heparan sulphate: an approach to measure chemical modifications on biological molecules with stable isotope labelling. *Biochem. J.,* **2005**, *389*(Pt 2), 383-388.
[http://dx.doi.org/10.1042/BJ20041827] [PMID: 15743272]

[59]   Oda, Y.; Huang, K.; Cross, F.R.; Cowburn, D.; Chait, B.T. Accurate quantitation of protein expression and site-specific phosphorylation. *Proc. Natl. Acad. Sci. USA,* **1999**, *96*(12), 6591-6596.
[http://dx.doi.org/10.1073/pnas.96.12.6591] [PMID: 10359756]

[60]   a) Krijgsveld, J.; Ketting, R.F.; Mahmoudi, T.; Johansen, J.; Artal-Sanz, M.; Verrijzer, C.P.; Plasterk, R.H.; Heck, A.J. Metabolic labeling of C. elegans and D. melanogaster for quantitative proteomics. *Nat. Biotechnol.,* **2003**, *21*(8), 927-931.
[http://dx.doi.org/10.1038/nbt848] [PMID: 12858183] b) Johannes, H.I.; Laurice, P. *in vivo* uniform N-15-isotope labelling of plants: Using the greenhouse for structural proteomics. *Proteomics,* **2004**, *1*(4), 226-234.

[61]   a) Hebeler, R. Study of early leaf senescence in Arabidopsis thaliana by quantitative proteomics using reciprocal N-14/N-15 Labeling and difference gel electrophoresis. *Mol. Cell. Proteomics,* **2008**, *1*(7), 108-120.
[http://dx.doi.org/10.1074/mcp.M700340-MCP200] b) Lanquar, V.; Kuhn, L.; Lelièvre, F.; Khafif, M.; Espagne, C.; Bruley, C.; Barbier-Brygoo, H.; Garin, J.; Thomine, S. 15N-metabolic labeling for comparative plasma membrane proteomics in Arabidopsis cells. *Proteomics,* **2007**, *7*(5), 750-754.
[http://dx.doi.org/10.1002/pmic.200600791] [PMID: 17285564] c) Kline, K.G.; Barrett-Wilt, G.A.; Sussman, M.R. In planta changes in protein phosphorylation induced by the plant hormone abscisic acid. *Proc. Natl. Acad. Sci. USA,* **2010**, *107*(36), 15986-15991.
[http://dx.doi.org/10.1073/pnas.1007879107] [PMID: 20733066]

[62]   Beynon, R.J.; Pratt, J.M. Metabolic labeling of proteins for proteomics. *Mol. Cell. Proteomics,* **2005**, *4*(7), 857-872.
[http://dx.doi.org/10.1074/mcp.R400010-MCP200] [PMID: 15849272]

[63]   Ong, S.E.; Blagoev, B.; Kratchmarova, I.; Kristensen, D.B.; Steen, H.; Pandey, A.; Mann, M. Stable isotope labeling by amino acids in cell culture, SILAC, as a simple and accurate approach to expression proteomics. *Mol. Cell. Proteomics,* **2002**, *1*(5), 376-386.
[http://dx.doi.org/10.1074/mcp.M200025-MCP200] [PMID: 12118079]

[64]   Zhang, R.; Sioma, C.S.; Thompson, R.A.; Xiong, L.; Regnier, F.E. Controlling deuterium isotope effects in comparative proteomics. *Anal. Chem.,* **2002**, *74*(15), 3662-3669.
[http://dx.doi.org/10.1021/ac025614w] [PMID: 12175151]

[65]   Yao, X.; Freas, A.; Ramirez, J.; Demirev, P.A.; Fenselau, C. Proteolytic ^{18}O labeling for comparative proteomics: model studies with two serotypes of adenovirus. *Anal. Chem.,* **2001**, *73*(13), 2836-2842.
[http://dx.doi.org/10.1021/ac001404c] [PMID: 11467524]

[66]   Schnölzer, M.; Jedrzejewski, P.; Lehmann, W.D. Protease-catalyzed incorporation of [18]O into peptide fragments and its application for protein sequencing by electrospray and matrix-assisted laser desorption/ionization mass spectrometry. *Electrophoresis,* **1996**, *17*(5), 945-953.
[http://dx.doi.org/10.1002/elps.1150170517] [PMID: 8783021]

[67]   Sha, X.M.; Hu, Z.Z.; Tu, Z.C.; Zhang, L.Z. Quantitative Determination of Gelatin in Ejiao by [18]O Labeling Combined with High Performance Liquid Chromatography-High Resolution Mass Spectrometry. *Shipin Kexue,* **2018**, *39*(12), 288-294.

[68]   a) Zhang, G.F.; Liu, T.; Wang, Q.; Chen, L.; Lei, J.; Luo, J.; Ma, G.; Su, Z. Mass spectrometric detection of marker peptides in tryptic digests of gelatin: A new method to differentiate between bovine and porcine gelatin. *Food Hydrocoll.,* **2009**, *23*(7), 2001-2007.
[http://dx.doi.org/10.1016/j.foodhyd.2009.03.010] b) Zhang, Q.; Tu, Z.; Wang, H.; Huang, X.; Shi, Y.; Sha, X.; Xiao, H. Improved glycation after ultrasonic pretreatment revealed by high-performance liquid chromatography-linear ion trap/Orbitrap high-resolution mass spectrometry. *J. Agric. Food Chem.,* **2014**, *62*(12), 2522-2530.
[http://dx.doi.org/10.1021/jf5002765] [PMID: 24606342]

[69]   Staes, A.; Demol, H.; Van Damme, J.; Martens, L.; Vandekerckhove, J.; Gevaert, K. Global differential non-gel proteomics by quantitative and stable labeling of tryptic peptides with oxygen-18. *J. Proteome Res.,* **2004**, *3*(4), 786-791.
[http://dx.doi.org/10.1021/pr049956p] [PMID: 15359732]

[70]   Petritis, B.O.; Qian, W.J.; Camp, D.G., II; Smith, R.D. A simple procedure for effective quenching of trypsin activity and prevention of ^{18}O-labeling back-exchange. *J. Proteome Res.,* **2009**, *8*(5), 2157-2163.
[http://dx.doi.org/10.1021/pr800971w] [PMID: 19222237]

[71]   Desiderio, D.M.; Kai, M. Preparation of stable isotope-incorporated peptide internal standards for field desorption mass spectrometry quantification of peptides in biologic tissue. *Biomed. Mass Spectrom.,* **1983**, *10*(8), 471-479.
[http://dx.doi.org/10.1002/bms.1200100806] [PMID: 6616023]

[72]   Gerber, S.A.; Rush, J.; Stemman, O.; Kirschner, M.W.; Gygi, S.P. Absolute quantification of proteins and phosphoproteins from cell lysates by tandem MS. *Proc. Natl. Acad. Sci. USA,* **2003**, *100*(12), 6940-6945.
[http://dx.doi.org/10.1073/pnas.0832254100] [PMID: 12771378]

[73]   Bantscheff, M.; Schirle, M.; Sweetman, G.; Rick, J.; Kuster, B. Quantitative mass spectrometry in proteomics: a critical review. *Anal. Bioanal. Chem.,* **2007**, *389*(4), 1017-1031.
[http://dx.doi.org/10.1007/s00216-007-1486-6] [PMID: 17668192]

[74]   Picotti, P.; Aebersold, R. Selected reaction monitoring-based proteomics: workflows, potential, pitfalls and future directions. *Nat. Methods,* **2012**, *9*(6), 555-566.
[http://dx.doi.org/10.1038/nmeth.2015] [PMID: 22669653]

[75]   Yao, X.D.; Mcshane, A.J.; Castillo, M.J. *Quantitative Proteomics in Development of Disease Protein Biomarkers*; Proteomic & Metabolomic Approaches to Biomarker Discovery, **2013**, pp. 259-278.
[http://dx.doi.org/10.1016/B978-0-12-394446-7.00017-0]

[76]   Kirkpatrick, D.S.; Gerber, S.A.; Gygi, S.P. The absolute quantification strategy: a general procedure for the quantification of proteins and post-translational modifications. *Methods,* **2005**, *35*(3), 265-273.
[http://dx.doi.org/10.1016/j.ymeth.2004.08.018] [PMID: 15722223]

[77]   Glickman, M.H.; Ciechanover, A. The ubiquitin-proteasome proteolytic pathway: destruction for the sake of construction. *Physiol. Rev.,* **2002**, *82*(2), 373-428.
[http://dx.doi.org/10.1152/physrev.00027.2001] [PMID: 11917093]

[78]   Nguyen, T.T.T.N.; Mistarz, U.H.; Costa, N.; Herbet, A.; Boquet, D.; Becher, F.; Rand, K.D. Investigating the utility of minimized sample preparation and high-resolution mass spectrometry for quantification of monoclonal antibody drugs. *J. Pharm. Biomed. Anal.,* **2018**, *159*, 384-392.

[http://dx.doi.org/10.1016/j.jpba.2018.07.012] [PMID: 30071466]

[79] Gomez-Zepeda, D.; Taghi, M.; Smirnova, M.; Sergent, P.; Liu, W.Q.; Chhuon, C.; Vidal, M.; Picard, M.; Thioulouse, E.; Broutin, I.; Guerrera, I.C.; Scherrmann, J.M.; Parmentier, Y.; Decleves, X.; Menet, M.C. LC-MS/MS-based quantification of efflux transporter proteins at the BBB. *J. Pharm. Biomed. Anal.,* **2019**, *164*, 496-508.
[http://dx.doi.org/10.1016/j.jpba.2018.11.013] [PMID: 30453156]

[80] Wong, J.W.H.; Cagney, G. An overview of label-free quantitation methods in proteomics by mass spectrometry. *Methods Mol. Biol.,* **2010**, *604*, 273-283.
[http://dx.doi.org/10.1007/978-1-60761-444-9_18] [PMID: 20013377]

[81] Yu, F.; Meza, J. *Design and Statistical Analysis of Mass Spectrometry-Based Quantitative Proteomics Data*; Proteomic Profiling & Analytical Chemistry, **2013**, pp. 179-204.
[http://dx.doi.org/10.1016/B978-0-444-59378-8.00010-4]

[82] Sun, Q.; Zhang, N.; Wang, J.; Cao, Y.; Li, X.; Zhang, H.; Zhang, L.; Tan, D-X.; Guo, Y-D. A label-free differential proteomics analysis reveals the effect of melatonin on promoting fruit ripening and anthocyanin accumulation upon postharvest in tomato. *J. Pineal Res.,* **2016**, *61*(2), 138-153.
[http://dx.doi.org/10.1111/jpi.12315] [PMID: 26820691]

[83] Pan, X.; Zhu, B.; Zhu, H.; Chen, Y.; Tian, H.; Luo, Y.; Fu, D. iTRAQ protein profile analysis of tomato green-ripe mutant reveals new aspects critical for fruit ripening. *J. Proteome Res.,* **2014**, *13*(4), 1979-1993.
[http://dx.doi.org/10.1021/pr401091n] [PMID: 24588624]

[84] Wang, Y.D.; Wang, X.; Ngai, S.M.; Wong, Y.S. Comparative proteomics analysis of selenium responses in selenium-enriched rice grains. *J. Proteome Res.,* **2013**, *12*(2), 808-820.
[http://dx.doi.org/10.1021/pr300878y] [PMID: 23244200]

[85] Old, W.M.; Meyer-Arendt, K.; Aveline-Wolf, L.; Pierce, K.G.; Mendoza, A.; Sevinsky, J.R.; Resing, K.A.; Ahn, N.G. Comparison of label-free methods for quantifying human proteins by shotgun proteomics. *Mol. Cell. Proteomics,* **2005**, *4*(10), 1487-1502.
[http://dx.doi.org/10.1074/mcp.M500084-MCP200] [PMID: 15979981]

[86] Wienkoop, S.; Larrainzar, E.; Niemann, M.; Gonzalez, E.M.; Lehmann, U.; Weckwerth, W. Stable isotope-free quantitative shotgun proteomics combined with sample pattern recognition for rapid diagnostics. *J. Sep. Sci.,* **2006**, *29*(18), 2793-2801.
[http://dx.doi.org/10.1002/jssc.200600290] [PMID: 17305241]

[87] Zybailov, B.; Coleman, M.K.; Florens, L.; Washburn, M.P. Correlation of relative abundance ratios derived from peptide ion chromatograms and spectrum counting for quantitative proteomic analysis using stable isotope labeling. *Anal. Chem.,* **2005**, *77*(19), 6218-6224.
[http://dx.doi.org/10.1021/ac050846r] [PMID: 16194081]

[88] Hoehenwarter, W.; van Dongen, J.T.; Wienkoop, S.; Steinfath, M.; Hummel, J.; Erban, A.; Sulpice, R.; Regierer, B.; Kopka, J.; Geigenberger, P.; Weckwerth, W. A rapid approach for phenotype-screening and database independent detection of cSNP/protein polymorphism using mass accuracy precursor alignment. *Proteomics,* **2008**, *8*(20), 4214-4225.
[http://dx.doi.org/10.1002/pmic.200701047] [PMID: 18924179]

[89] a) Kito, K.; Ito, T. Mass spectrometry-based approaches toward absolute quantitative proteomics. *Curr. Genomics,* **2008**, *9*(4), 263-274.
[http://dx.doi.org/10.2174/138920208784533647] [PMID: 19452043] b) Kalra, H.; Adda, C.G.; Liem, M.; Ang, C-S.; Mechler, A.; Simpson, R.J.; Hulett, M.D.; Mathivanan, S. Comparative proteomics evaluation of plasma exosome isolation techniques and assessment of the stability of exosomes in normal human blood plasma. *Proteomics,* **2013**, *13*(22), 3354-3364.
[http://dx.doi.org/10.1002/pmic.201300282] [PMID: 24115447]

[90] Abdallah, C.; Dumas-Gaudot, E.; Renaut, J.; Sergeant, K. Gel-based and gel-free quantitative proteomics approaches at a glance. *Int. J. Plant Genomics,* **2012**, *2012*(10)494572

[http://dx.doi.org/10.1155/2012/494572] [PMID: 23213324]

[91]   a) Yan, W.; Chen, S.S. Mass spectrometry-based quantitative proteomic profiling. *Brief. Funct. Genomics Proteomics,* **2005**, *4*(1), 27-38.
[http://dx.doi.org/10.1093/bfgp/4.1.27] [PMID: 15975262] b) Wang, M.; You, J.; Bemis, K.G.; Tegeler, T.J.; Brown, D.P.G. Label-free mass spectrometry-based protein quantification technologies in proteomic analysis. *Brief. Funct. Genomics Proteomics,* **2008**, *7*(5), 329-339.
[http://dx.doi.org/10.1093/bfgp/eln031] [PMID: 18579615] c) Megger, D.A.; Bracht, T.; Meyer, H.E.; Sitek, B. Label-free quantification in clinical proteomics. *Biochim. Biophys. Acta,* **2013**, *1834*(8), 1581-1590.
[http://dx.doi.org/10.1016/j.bbapap.2013.04.001] [PMID: 23567906]

[92]   Wolf-Yadlin, A.; Hautaniemi, S.; Lauffenburger, D.A.; White, F.M. Multiple reaction monitoring for robust quantitative proteomic analysis of cellular signaling networks. *Proc. Natl. Acad. Sci. USA,* **2007**, *104*(14), 5860-5865.
[http://dx.doi.org/10.1073/pnas.0608638104] [PMID: 17389395]

CHAPTER 3

Chemometrics as a Powerful and Complementary Tool for Mass Spectrometry Applications in Life Sciences

Yahya Izadmanesh[1,*] and **Jahan B. Ghasemi**[1,*]

[1] *Faculty of Chemistry, University of Tehran, Tehran, Iran*

Abstract: Because of its unique capabilities, mass spectrometry is an indispensable part of life science research. In this chapter, a review is made on aids of chemometrics in life sciences applications of mass spectrometry. Because of the increasing complexity of biological samples and ongoing technological enhancements of mass spectrometers, huge sum of data are provided for each biological sample. If the routine exploratory tools are used for data exploration, much of the information is not extractable and hence it gets lost. However, chemometrics helps to explore data thoroughly and extract maximum amount of information. The most common aids of chemometrics in bio-based mass spectrometry data is for experimental design, noise reduction, classification, library search, identification of biomolecules, finding the biomarkers, data compression and data mining.

This chapter is focused on the different aspects of using chemometrics for the analysis of mass spectrometry data in omics and biomedical images. In the first part, chemometrics applications for mass data in omics sciences (metabolomics and proteomics) are revealed. The mass data in omics are mainly provided by hyphenation of mass spectrometry with chromatographic techniques, i.e., gas chromatography (GC), liquid chromatography (LC) and electrophoretic techniques. In the second part of the chapter, the benefits of using chemometrics for mass spectrometry images are revealed. The data of these images are gathered by mass spectrometer itself or hyphenation with chromatographic techniques. Since, hyphenated methods are used for both omics and biomedical imaging, some of the chemometrics methodologies used in these two disciplines may be the same.

Keywords: Biomolecules, Biomarker detection, Biomedical imaging, Chemometrics, Data analysis, Data binning, Data compression, Data mining, Experimental design, Genomics, Life science, Mass spectrometry, Metabolomics, Multivariate curve resolution-alternating least squares (MCR-LAS), Multivariate

---

* **Corresponding author Yahya Izadmanesh:** Faculty of Chemistry, University of Tehran, Tehran, Iran; Tel: (+98) 2161112726; Fax:(+98) 2166972074; E-mails: yizadmanesh@alumni.ut.ac.ir and Jahan.ghasemi@ut.ac.ir

Atta-ur-Rahman, M. Iqbal Choudhary & Syed Ghulam Musharraf (Eds.)
All rights reserved-© 2020 Bentham Science Publishers

methods, OMICS, Proteomics, Regions of Interest (ROI), Statistics, Variable selection, XCMS software.

INTRODUCTION

Mass spectrometry (MS) records the mass-to-charge ratio (m/z), of the sample components. In a simplified scheme, mass spectrometer consists of three parts: the ionization source, the mass analyzer, and detector. Therefore, the components must get ionized in the ionization source before m/z measurement. The ionized components are then transferred to the mass analyzer that separates the components according to m/z of the ions. Then, the detector detects the separated components, and mass spectrum is recorded. A mass spectrum is depicted as a diagram which m/z in x-axis and the intensity of the signal for is in y-axis [1]. In recent mass spectrometers carrying Orbitrap or FT-ICR analyzers, the mass analyzer itself performs the mass analysis and detection. There is not a separate detector where the ions will hit. Nowadays, most of life science studies use high resolution mass spectrometry in Orbitrap equipment [2].

For many years, mass spectrometry has played an important role in the health and life sciences researches. Investigation of complex biomolecules by mass spectrometry approaches is crucial in molecular life science research. In order to monitor qualitative and quantitative changes within hundreds or thousands of biologically active components, including proteins/peptides, lipids and metabolites mass spectrometry is an indispensable part. The mass spectrometry aided studies can help to understand pathophysiology of disease development at a molecular level and to monitor the effect of pharmacological treatment [3].

Often, the scientific measurements can be collected in a data matrix, where each row constitutes an observation and the columns represent the measured variables or factors (e.g., wavelength, mass number, chemical shift). This type of data collection generates huge data tables, which are hard to extract relevant information without appropriate tools. The power of chemometrics becomes relevant when working with these data which leads to the extraction of plentiful of useful information. However, in biology, chemometrics has been largely neglected in favor of traditional statistical methods. The overwhelming size and complexity of the data has forced biologists toward the use of multivariate statistical methods such as robust modeling [4 - 7].

Chemometrics is an interdisciplinary science that uses mathematics, statistics and computer sciences to overcome the huge data challenges. The interpretation of chemical data by chemometric methods helps to design, select and optimize experiments, and extract relevant information. Chemometrics can be defined as computer applications in chemistry, including data acquisition and processing, optimization, intelligent laboratory systems, robotics, statistics, pattern recognition, cluster analysis, library search, structure property relationship, modeling information theory, artificial intelligence and expert systems [8, 9]. The computer hardware and software have widespread use in mass spectrometry [10, 11]. Conversely, the challenging problems from mass spectrometry has stimulated development of chemometrics. The chemometrics is a powerful tool for data acquisition, data handling, instrument control, and data interpretation in mass spectrometry [12].

In this chapter, we are going to have a brief survey of adaptation of different chemometric methods into the mass spectrometry and mass data collected from biological samples. This chapter will provide an overview on how chemometrics has evolved as an invaluable and indispensable part of mass spectrometry in life sciences. However, it should be reminded that this chapter may not cover all data analysis and computing approaches used in the life sciences mass spectrometry data.

MASS SPECTROMETRY OMICS DATA

OMICS can be defined as the study of the abundance and (or) structural characterization of biomolecules in living organisms (see Fig. **1**). The high-throughput omic technologies are useful for clinical characterization of diseases in organisms and evaluation of efficacy of existing or under-development therapies [13].

OMICS include DNA studies (genomics and epigenomics), RNA studies (transcriptomics), proteins study (proteomics), and metabolites studies (metabolomics). Recently, another omic subdiscipline called fluxomics, is developed to study the total set of fluxes in the metabolic network (fluxome) of the bio organisms. Other omic platforms include lipidomics, glycomics, foodomics, interactomics, and metallomics [14].

The technologically improved analytical techniques produce large omic data sets with complex structures. In mass spectrometry based analytical methods (GC-MS, LC-MS, GC×GC-MS, LC×LC-MS and electrophoresis-MS), highly complex OMIC data, lead to exploratory and interpretative challenges. To get useful information on major events taking place in an investigated system, data need to be processed and analyzed. The multivariate data are highly complex, and data analytical techniques must be used to cope with data challenges, including noise, collinearities, and missing data [15].

As indicated in Fig. (**1**) mass spectrometry is mostly used in metabolomics and proteomics. Hence, in this chapter, the data analysis strategies will be mostly highlighted for metabolomics and proteomics. However, in the case of genomics, transcriptomics, or other sub-disciplines, principals and methodologies will be the same.

Fig. (1). Overview of OMIC platforms: target molecules, used analytical methodologies and structure of the generated data (GE N°: number of genes, δ: chemical shift, *m/z*: mass-to-charge ratio, rt: retention time, I: intensity). Reprinted with permission from Ref [13].

METABOLOMIC MASS DATA

In general terms, metabolomics studies are classified to targeted and untargeted metabolomics. The reason for this classification is the generation of different types of data in these two approaches, which require being handled differently. In targeted studies [16], the metabolites are known, whereas in untargeted studies, the metabolites are unknown and the goal is to identify as much as metabolites possible [17 - 21]. In untargeted approach, the number of compounds analyzed is larger than targeted approach and entire metabolite signals must be processed. Among the signals, few are finally identified as candidate biomarkers [22]. There are plenty of data analysis strategies in the literature, which makes data analysis an open and challenging task.

MS-based metabolomics is a rather young filed, and new concepts are being regularly published or updated [23, 24], because high-resolution mass spectrometry provides noisy and collinear high-dimensional data structures. The data require extensive preprocessing to assess sample classification/discrimination and biomarker discovery [25]. Thus, for comprehensive metabolomics, more advanced chemometric methods are in need [26]. Fig. (**2**) shows the general scheme of a typical untargeted mass based metabolomics study using capillary electrophoresis (CE-MS) for biomarkers identifications of baker's Yeast [26]. In a typical study, the data are collected, preprocessed, data mined and modeled by chemometric methods to find the biomarkers. However, the data analysis algorithms and methods used in each step may differ from study to study.

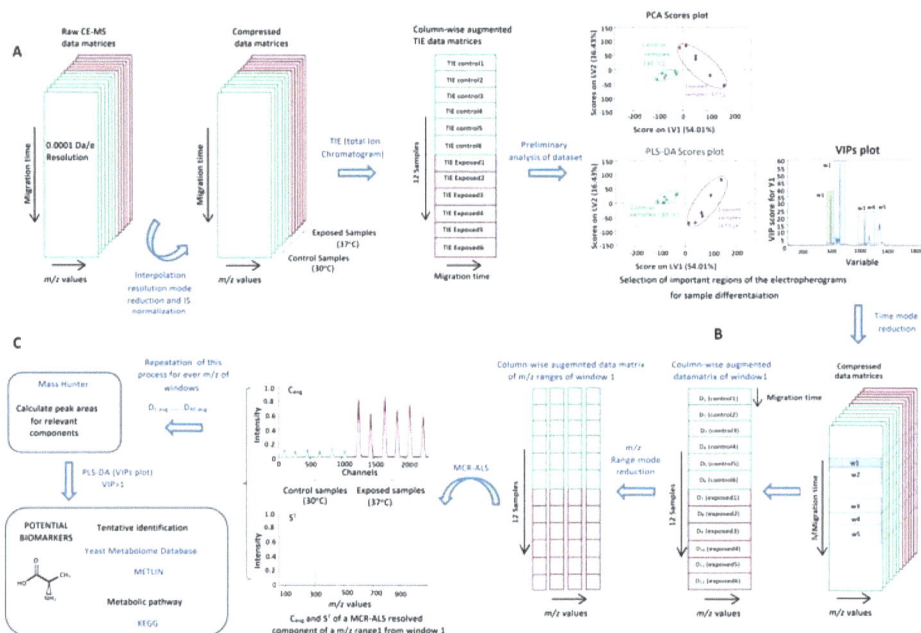

Fig. (2). Workflow of CE-MS data analysis: (**A**) Pre-processing and preliminary data analysis, (**B**) data arrangement and Chemometric modeling. (**C**) Tentative identification of relevant metabolites and metabolic pathways. Reprinted with permission of Ref [26]. The chemometric algorithm used in this study is MCR-ALS.

Chemometrics presents a complete theory and methodology for every step of metabolomics, i.e., sampling, experiment design, data pre-processing, metabolite identification, variable selection, and modeling [27, 28]. It should be noted that classification of chemometrics applications presented here may be applicable to other omic branches. The main areas of applications of chemometrics in metabolomics mass data are discussed below:

Experimental Design

The ultimate goal of experimental design is exploration of the maximum amount of information in the fewest number of experimental runs. The basic idea is to design a small set of experiments (usually 10-20 experiments), in which all relevant factors are varied systematically [29 - 31]. Addition of extra experiments helps to investigate factors more thoroughly. Experimental design helps to minimize the noise level by means of averaging, efficiently mapping the functional space, and exploring interactions and synergisms. The design of scientific studies is important for robust scientific conclusions. Experimental design ensures that biological variations are significantly greater than process variations. Without experimental design, large collected datasets provide no relevance to the biological objectives and can lead to false observations and conclusions about organisms.

Experimental design is the first step in the metabolomics workflow and has high influence on metabolite content and biological interpretation of data. Thus, a good experimental design and the appropriate choice of a sample processing method are prerequisites for success of any metabolomics study [32, 33]. All of the available experimental design methodologies can be used dependent on the user and system under investigation. The factorial design methods can be used to screen the important and not important factors in the sample preparation step. The response surface methodology methods can be used to optimize the level of important factors in order to identify biomarkers and determine their concentration changes at highest possible accuracy and precision.

Data Preprocessing

Raw data must be processed to generate list of metabolites [23]. The first step is elimination of the data variance and bias to reduce the complexity and enhance significant signals [34]. To this end, several algorithms have been developed and multiple open source programs have been applied to process raw MS data [34 - 41]. Usually, the algorithms and data are exchanged within the community. In general, data preprocessing workflow consists of four basic modules: noise filtering and baseline correction, peak detection and deconvolution, alignment, and normalization.

Noise Filtering and Baseline Correction

Noise filtering separates background signals originated from the chemical matrix, instrumental interference, measurement noise or baseline distortions [42]. For baseline correction of one-way data (e.g. a mass spectrum), the two terminals of a signal are manually selected and fitted by a piecewise linear approximation to calculate a curve as a baseline [43]. This procedure is exhausting, and its accuracy is highly user-dependent, and hence, other alternative algorithms have been developed. Two newer algorithms in this context are automatic two-side exponential baseline correction algorithm (ATEB) [44] and adaptive iteratively reweighted penalized least squares (airPLS) [43]. These approaches automatically remove the nonlinear and linear baselines. Unlike former methods, these algorithms are very fast, robust, and do not require prior knowledge.

LC-MS noise filtering is far more complicated than GC-MS, because both chemical and random noises are included in the former. Chemical noise is generated by buffer molecules and solvents and are more intense at the beginning and end of the elution [45]. This noise causes a shift in the baseline at the intermediate mass range of LC-MS spectra which has led to the development of two noise filtering approaches. In the first approach, the spectrum is segmented and the baseline is fitted by a linear regression [46]. In the second approach, the baseline is removed by estimating the background of a two dimensional intensity image [47].

Peak Detection and Deconvolution

The goal of peak detection and deconvolution is identification and quantification of metabolite signals in a sample [25, 48]. These procedures are used for alignment and identification of biomarkers, and reducing the data complexity [49]. However, because of the complexity of the samples and multiple sources of noise in data, automatic discrimination of the noise from signal is truly challenging. Definition of threshold to identify noise and a signal is difficult, especially when peaks have low-response values. High intensity response is not always indicator of real peaks, because some sources of noise produce intense signals. Conversely, low peaks may correspond to real signals. To overcome these challenges, constraints are applied on the shapes, intensity, area and signal to noise of peaks [13, 15, 25, 26, 49 - 53].

Normalization

Normalization removes confounding variations related to experimental sources, and keeps variations indicating biological events [23]. Normalization can be done by calculation of the relative ratio of the abundance of analytes peak to all of the peaks, if metabolites signals are stable [54]. However, this assumption is highly unlikely because of systematic laboratory errors and differences among large scale biological experiments, therefore, normalizing is always an important challenging step [55, 56].

Metabolites Identification

Reliable identification of metabolites from MS spectral data is challenging especially in untargeted analysis, because the metabolites are biochemically diverse. High-throughput metabolite identification with reasonable accuracy is achievable owing to the development of powerful computation methods, advances in mass spectrometry instrumentation, wealth of knowledge on ion fragmentation, and well-established databases [57 - 60].

Identification using GC-MS data is done by comparison of experimental MS spectra with the reference in the mass spectral library. The suggested compound with the highest similarity score is considered as the identified compound [61, 62]. However, because of existence of isomers and co-eluted components, usually compound with the highest similarity score isn't the real target compound, therefore, careful extra checking such as calculation of Kovat's retention index (RI) of target is necessary [63, 64]. Another identification methods that work independent of mass spectral library learn the structural features of unknown compound from experimental mass spectra and then deduce structure by comparison of features to previously constructed learning models [65 - 67].

Identification procedure in LC-MS data is different than GC-MS data, due to higher experimental variations [68]. The identification is based on two aspects of LC-MS data: accurate mass with other information like isotopic distribution and MS/MS spectra. The accurate mass determination is the first step [58]. Then, the formula generation method or the searching in a large compound database or metabolism network is used. All combinations of predefined elements with constraints of element number and mass range are executed to generate a formula. This approach generates a large number of candidate formulas, especially for high molecular mass compounds, and hence, it is impracticable to obtain a single assignment of formula to each m/z. To overcome this problem, more constraints are applied to limit number of final candidates. Similarity checking of isotopic distribution is one of the best indicators and with this checking majority of the fake formulas get rejected [69, 70]. The formula candidates are ranked by comparison of the instrument-determined isotopic distribution to the simulated one, in a way that the top ones are the most similar [71, 72]. This strategy is proved to be reliable for high-throughput metabolomics analysis with higher-resolution instruments. After formulas determination, decoding them to known metabolites in LC-MS feature annotation is performed, typically by searching large chemical substance databases [73, 74].

Multistage mass spectrometers (MS^n) are highly effective for structure exploration. The LC-MS systems are usually equipped with MS^2. The ionized species in the specified m/z range are gradually dissociated into charged and neutral pieces and all the charged fragments and precursor ions are recorded as MS^2 spectrum. The experimental conditions (e.g., collision energy) in MS^n systems are not as standardized like GC-MS and the sizes of currently constructed MS^n libraries are much smaller compared to the number of metabolites in a typical metabolism [75, 76]. Thus, metabolite identification by spectral library search is not the best option in MS^n. Alternatively, the focus is on developing computational methods to interpret MS^n spectra without using spectral libraries. The computational MS^n identification algorithms can be categorized into three basic approaches: mass spectrum prediction, *in silico* fragmentation, and de novo elucidation [77]. Mass spectrum prediction is mainly used for MS^2 data, and gives satisfying results in electrospray ionization (EI) spectrum interpretation and also is the basic module for peptide identification in hypothesis-driven proteomics. The diversity of small compounds is the main challenge for accurate MS^2 spectral prediction using this approach. For prediction of the MS^2 spectrum of a structure, all possible reactions that occur during the fragmentation of a structure are extracted from its own fragmentation reaction library [78, 79].

In silico fragmentation selects structure among all candidates that best explains the MS^2 spectrum [80 - 84].

In *de novo* method, first the formulas of fragments are determined using high resolution *m/z* and then the structure of a precursor ion are identified using these formulas and the known previously fragmentation pathways that generate these ions. The most appropriate flowchart employed in this method is the construction of a fragmentation tree with nodes being fragment formulas, edges being neutral losses, and the root being precursor [85 - 89].

Resolution Algorithms

XCMS is a favorite method among the metabolomic community for feature detection, and it is being used for a broad range of applications. XCMS is designed for chromatographic feature detection and automatic processing of huge size full scan LC-MS data and identifies candidate metabolites by combination of peak detection and retention time correction algorithms and methods. For each proposed candidate, XCMS gives *p*-value and fold change. *P*-value is a statistical parameter comparing the integrated peak areas of this candidate in control versus treated samples and fold change is the ratio of the integrated peak areas of the treated samples versus the control samples [90, 91].

MZmine is as an open source software for LC-MS data processing [92]. The first version of MZmine was developed for data processing and visualization. A weakness of MZmine was limited ability of synchronization with newer algorithms developed in the scientific community, thus the new version MZmine 2 was released using Java programing to overcome the synchronization problem [93]. It consists of modules for raw data import, raw data processing, peak detection, peak list alignment, peak identification, visualization, normalization and statistical analysis.

MCR-ALS is a popular chemometric method used for resolution of compounds mass spectrums and resolution of overlapped chromatographic peaks [51]. MCR-ALS has been proposed as an alternative approach to detect potential biomarkers in untargeted metabolomics studies [25, 50, 53, 94]. MCR-ALS decomposes the experimental data into their chromatographic elution profiles and to the mass spectra of each resolved component. The main difference between XCMS and MCR-ALS lies in peak detection and resolution. While each feature in XCMS is identified by retention time and a unique *m/z* value, features in MCR-ALS are mathematical concepts that are characterized by their elution profiles and mass spectra that more than one possible MS feature can be assigned to a elution profile [94, 95].

Other softwares are also developed for data processing of metabolomic and proteomic LC-MS dataset such as Trans Proteomic Pipeline [96], Trequips [97], OpenMSTOPP [98], and ProteoWizard [99]. These methods are mostly command-line oriented with fixed features having limited applicability.

Variable Selection

The aim of variable selection is extraction of biomarkers from all of the detected metabolites. Statistically speaking, this is an optimization approach that discovers biomarkers from large number of metabolites [100]. Some of the strategies for biomarker detection are based on statistical features of variables (metabolites), whereas others are based on the optimization algorithm. Variable ranking and variable subset selection are two strategies used in this context [101].

Variable Ranking

Variable ranking is mostly used for identification of biomarkers. Variable ranking assigns a measure of importance to each variable on the basis of predefined criteria. PLS-based criteria are the most frequently used [102] and include PLS loading weights (LW) [6], variable importance on projection (VIP) scores [103], regression coefficient (RC) [7], target projection (TP) [104], and selectivity ratio (SR) [105], with VIP being the most popular criteria in metabolomics [106]. The efficiency of these methods is data-dependent, and each method has its own pros and cons [107, 108]. Different variable ranking results are generated using each of variable ranking methods because of different principle and assumptions. To unify the results of different methods, rank aggregation method has been developed [109].

Variable Subset Selection

The aim of subset selection is to search and find an optimal subset representative of all variables. Variable ranking methods can be transformed into a variable subset selection algorithm by applying a subjective or statistical threshold on the variable importance values [110]. Usually, a balance between prediction accuracy of model and the number of selected variables is considered. For this purpose, the best approach is cross validation (CV) procedure. In CV procedure, after ranking variables from the most important to the least, models are built by adding these variables sequentially until all are included, and CV error belonging to each model is recorded. The best variable subset is the first n variables if minimum CV error is achieved after adding n^{th} variable [111]. Variable subset selection generally achieves better prediction accuracy than variable ranking because the latter considers the specific interactions between the classifier and dataset. In the process, subset selection utilizes a re-sampling mechanism to avoid over-fitting.

The collective effect of variables and variable interactions should be considered because the performance of variable selection becomes better [112, 113]. Several methods such as INTERACT [113], model population analysis (MPA) [114], Subwindow permutation analysis (SPA) [115], margin influence analysis (MIA) [116], binary matrix sampling (BMS) [117], random variable combination (VIAVC) [107], variable iterative space shrinkage approach (VISSA) [118] and variable complementary network (VCN) [119] are suggested for studying interaction effects. Table **1** lists variable selection methods, their classifiers, algorithms, computation speed, and classifications. The references for each method are also given for readers interested in further details of the methods. However, the figures of merits are dependent to the case study and they may vary according to case study. As a general statistical rule, as the number of samples increases, the figures of merits of these statistical approaches improve. But there must be a balance between number of samples and figures of merits, because indefinite number of samples is not practical and does not happen in real word.

Table 1. Classification of variable selection methods. Reused and rearranged with permission from ref [44].

Method	Classifier	Algorithm	Interaction	Computation Speed	Type	Ref. No
PLS Weights	PLS	PLS loading weight matrices are used.	NO	High	Ranking	[6]
PLS-VIP	PLS	Summation of the importance of variables using loading weights of PLS latent variables are calculated.	NO	High	Ranking	[103]

(Table 1) cont.....

Method	Classifier	Algorithm	Interaction	Computation Speed	Type	Ref. No
PLS regression coefficient	PLS	A measure of association between variables and the responses are used.	NO	High	Ranking	[7]
Correlation	No classifier	The correlation between variables and classification label are calculated.	NO	High	Ranking	[120]
Information gain	No classifier	-	NO	High	Ranking	[121]
Euclidean distance	No classifier	-	NO	High	Ranking	[122]
Mutual information	No classifier	-	No	High	Ranking	[123]
CARS	No classifier	Uses feature selection based on the regression coefficients.	No	High	Ranking	[124]
GA-PLS-DA	PLS-DA	GA is used as optimization algorithm to find the optimal subset using PLS-DA.	No	Low	Subset selection	[125]
PSO-SVM	SVM	PSO is used as optimization algorithm to find the optimal subset using SVM.	No	Medium	Subset selection	[126]
Random Forest	Decision three	Ranks the permuted randomly variables as a function of percent increase of misclassification error.	Yes	Medium	ranking	[113]
SPA	PLS-DA	Identification and ranking of the informative variables are performed by calculating difference between the prediction errors of normal and permutated sub-window of variables.	Yes	Medium	Ranking	[115]
MIA	SVM	Calculates a measure using the difference between the prediction errors after inclusion and exclusion of variables in the SVM model.	Yes	Medium	ranking	[116]
INTERACT	No classifier	The interaction features are explored using inconsistency and symmetrical uncertainty measurements.	Yes	High	Subset selection	[127]

(Table 1) cont.....

Method	Classifier	Algorithm	Interaction	Computation Speed	Type	Ref. No
VCN	PLS-DA	The complementary information between variables is computed and then the biomarkers are discovered using interaction features.	Yes	Medium	Ranking	[119]
IRIV	PLS	The optimal subset of variables is searched by measuring the difference between the prediction errors before and after inclusion of each variable in the model.	Yes	Medium	Subset selection	[109]
VISSA	PLS	The optimal variable combinations are searched by smooth shrinking of the variable space.	Yes	Subset selection	Medium	[118]

Modeling of the Data

Machine-learning methods are used to explore the high-dimensional metabolomics datasets and discover valuable information on biological events. The modeling methods can be categorized to supervised methods, unsupervised methods, and nonlinear modeling methods. Main performance characteristics of the machine learning methods are summarized in Table **2**. This table contains category, advantages and disadvantages of each method. The references for each method are also given for the readers interested in details of methods. The figures of merits of the given methods are data dependent as each of these methods have advantages and disadvantages.

Unsupervised Methods

Unsupervised methods are usually used for exploration of the dataset structure, finding trends and grouping in the dataset. The mostly used unsupervised methods in metabolomics include principal component analysis (PCA), hierarchical cluster analysis (HCA), self-organization mapping (SOM), and MCR-ALS. PCA transforms the high-dimensional variables into a small number of orthogonal factors, called PCs, which contain the systematic variance of the data [143]. PCA projects the samples into low dimensional (usually two- or three-dimensional) PC space, thus the visualization of the sample distribution is possible. HCA classifies relatively similar samples in one cluster and relatively dissimilar objects in another [144]. SOM, a neural-network algorithm, uses a non-linear projection of data on a lower dimensional grid [145], which the clustering in the data space and the metric-topological relations of the variables becomes visible. SOM is a useful tool for characterization of metabolic patterns and interrelationship between samples [133]. PARAFAC2 [146, 147], an improved version of PARAFAC [148], can be used to model three-way data with a trilinear structure. PARAFAC2 is the generalization of PCA to a higher order of data, which uses extra constraints to get real-world results for experimental data. The simultaneous processing of all samples, deconvolution of metabolites, elimination of chromatographic baseline, and alignment of retention time shifts can be done by this method. The ability of PARAFAC2 to find and model the shifted peaks of the same chemical compounds in different samples is an advantage. Its disadvantage is sensitivity to noise. MCR-ALS is based on a bilinear model, which decomposes the measured data data matrix, into a set of pure component contributions in two factor matrices. MCR-ALS has been successfully applied to modeling of omics data and has been shown to be very effective and powerful [25, 50, 51, 149]. MCR-ALS can handle peak shifts smoothly without the need for alignment. It is insensitive to noise and results are real-world.

Table 2. List of multivariate analysis methods used for unsupervised and supervised modeling. Reused and rearranged with permission from ref [44].

Method	Category	Advantage	Disadvantage	Ref.
PCA	Unsupervised	Provides an overview of large datasets.	Class information is not considered.	[128, 129]
HCA	Unsupervised	Provides an overview for the clusters of samples.	Class information is not considered. Variable importance is not calculated.	[130, 131]
SOM	Unsupervised	Can handle nonlinearity in the data.	Class information is not considered.	[132 - 134]
PARAFAC	Unsupervised	Can handle shifted data.	It is sensitive to noise.	[135, 136]
MCR-ALS	Unsupervised	Can handle shifted baseline in data. Constraints are more realistic, thus giving real results. Less sensitive to noise. It is fast compared to other modeling methods. Can be used for alignment of chromatographic peaks.	For big data it takes time to do analysis. This problem exists for all the modeling methods.	[51 - 53]
LDA	Supervised	Easy and fast. Suitable for linear and low dimensional data modeling.	Not suitable for modeling of high dimensional data.	[137, 138]
PLS-DA	Supervised	Particularly suitable to linear and co-linear data.	Not suitable for unbalanced data.	[139]
OPLS-DA	Supervised	Particularly suitable to linear and co-linear data.	Not suitable to unbalanced data.	[140]

(Table 2) cont.....

Method	Category	Advantage	Disadvantage	Ref.
SVM	Supervised	Suitable for flexible modeling of linear and nonlinear data.	The results aren't transparent and model tuning is challenging.	[141]
RF	Supervised	Applicable for linear and nonlinear models and can handle outliers.	The computation speed is relatively low.	[142]

Supervised Methods

Supervised techniques are used for modeling data having known structures. They are classified to linear methods, such as PLS-DA, linear discriminant analysis (LDA), orthogonal projections to latent structures discriminant analysis (OPLS-DA), and nonlinear methods, including RF and SVM. LDA finds a linear model from the original variables, which has maximum between-class variance and minimum within-class variance. PLS-DA is a combination of PLS regression and LDA and the most widely used supervised method for classification, [150]. PLS-DA can handle highly collinear data, provide excellent insights into the cause of discrimination by checking the behavior of variables and is useful for biomarker discovery. The OPLS-DA is modified version of PLS-DA [151] and the systematic variations in data is classified into two classes by orthogonal signal correction (OSC) technique: one class exhibits linear responsiveness, whereas another is linearly orthogonal to the response [152]. OPLS-DA provides better visualization and interpretation than PLS-DA and has been widely used for modeling and discovery of biomarker in metabolomics [153].

Non-linear Methods

Complex interactions taking place in biological organizations lead to nonlinear response in biological processes and hence the non-linear pattern recognition methods are required for modeling of metabolomics data. Kernel-PLS [154], support vector machines (SVMs) [155] and random forest (RF) [113] are three popular methods used as nonlinear methods in metabolomics. Kernel-based models transform data using some specific functions called kernels which transform the original non-linear data into a linear solvable data [156]. SVM is particularly suitable for small size data and is a kernel-based classifier that uses support vectors to define decision boundaries and separate binary classes. The disadvantages of SVM are the lack of model transparency, difficulty in calculation of variable importance, and not providing universal means of solving non-linear problems [157]. RF is an ensemble-learning method that provide large number of classification and regression trees and is highly powerful classifier for high-dimensional data [158].

Model Optimization and Validation

The parameters optimization is of great importance when constructing a model. CV is the most commonly used model optimization method [159]. Leave-one-out CV, K-fold CV and Monte Carlo CV are the most common subtypes of CV used for modeling optimization [160 - 163].

Model validation evaluates model prediction ability. Ideally, model validation uses an independent test set which is representative of training data and independent from these data. Several criteria are used to evaluate the prediction ability of a model such as sensitivity, specificity, accuracy, the receiver operating characteristic curve, and the cross-validated coefficient of determination relationships between groups [106].

PROTEOMIC MASS DATA

The aim of a proteomics experiment is studying the relationship between the protein expression and certain sample groups (*e.g.* distinct disease classes), or studying the relationships between proteins. The goal of the former is comparison of the protein expression profiles in two different biological samples. The goal of the latter is to group the proteins with respect to their expression profile in a certain biological sample [164 - 166]. Most proteomics experiments use high throughput technologies such as two dimensional electrophoresis, two dimensional chromatography, mass or protein arrays to collect the expression data of thousands of proteins simultaneously [167]. These experiments yield high dimensional datasets which statistical tools are essential for data analysis and prevention of false conclusions [168, 169].

Mass spectrometry proteomics is relatively young and evolving field, but it is gaining importance. Combination of innovative experimental strategies [170], and advanced computational methods [165, 171], has enabled global study of cellular proteomes by MS based proteomics. Relating complex proteomics data to biological processes is impossible without computer-aided data analysis strategies specially bioinformatics [172, 173]. Bioinformatics is a tool for functional analysis and data mining of data sets to obtain biologically interpretable results and insights.

In common procedure of MS-based proteomics, proteins are digested by enzymes to give peptides, which are subsequently analyzed online by LC-MS or LC-MS×MS techniques [174]. Protein identification from the resulting data is a multistep process. First, experimental mass spectrums of the peptides resulted from an unknown protein are compared with theoretical sequences from a sequence database. The goal is to find a protein in the sequence database that has the highest similarity with the experimental data. In practice, the acquired peak list is presented to a specialized software program to do the comparison. Software may be provided within the LC-MS, but there are also several freely available web based tools as well as commercial softwares that are designed for this task [167]. For precise identification, the spectra should contain clear peaks from the peptides of the protein sample, and ideally no interferents. The quality of MS spectra depend on sample preparation, sample matrix complexity, the figures of merit of MS instrument (accuracy, precision, resolution), and availability of calibration standards. Four general types of methods are used for identification of proteins in proteomics: (1) *de novo* sequencing [175], (2) the use of unambiguous "peptide sequence tags" for known sequences search [176], (3) cross correlation methods that correlate experimental spectra with theoretical spectra [177], and (4) probability-based matching that calculates a score based on the statistical significance of a match between an observed peptide fragment and those calculated from a sequence search library [178].

De novo sequencing identifies peptide sequence from an MS/MS spectrum without any known sequence knowledge. *De novo* sequencing is used for analyzing spectra that are not identifiable by other search methods, *i.e.,* no homologous sequences exist in the database. However, *de novo* sequencing followed by a database search is capable of competing with other three approaches.

In LC-MS based expressional proteomics, multiple samples from different groups are simultaneously analyzed and procedures are developed for peak quantification, peak alignment and data quality assurance. Theoretically, it is predicted that a biomolecule should have the same retention time, molecular weight and signal intensity, but, this assumptions are rarely real because of experimental variations [54]. Spectral deconvolution, peak alignment and data quality checking are crucial tasks to ensure robust data analysis in proteomics [179].

GENOMICS MASS DATA

Genomics is the systematic study of genes, their functions, and interactions. In the past decades, genomics and proteomics studies were focused on one gene or one protein at a time. With the developments of high-throughput biological and biotechnological technologies, dramatic changes have happened in this context. The genomic data are mainly obtained by microarray sequencing, and mass spectrometry has limited application [180]. However, when facing the challenges of data analysis of mass spectrometry data of genomics, the same methodologies mentioned for proteomics and metabolomics data applies here [181].

MASS SPECTROMETRY IMAGING

In recent years, the introduction of hyperspectral imaging techniques to carry out fast and relatively cheap analyses of multiple compounds spread over the surface of a sample has increased dramatically [182, 183]. The imaging have been used in many areas, such as food processing, food quality control [184], environmental applications and biomedical studies [185 - 187]. For these applications, complete spectrum (usually from a vibrational spectroscopic technique such as NIR, Raman or mass spectroscopy) of the sample is collected for pixels of the sample surface and further processed by data analysis techniques to collect useful information.

Mass Spectrometry Imaging (MSI), is invaluable for the study of real biological samples such as cells or tissues [188 - 190]. MSI has high chemical specificity toward biomolecules easing the way for simultaneous analysis of multiple molecular species in a very wide mass range, from small (i.e. metabolites) to large molecules (*i.e.* proteins). MSI provides information about the presence or absence of biomolecules, and their spatial distribution in the sample surface [191]. Integration of the spatial information with the chemical specificity based on the highly accurate mass data allows unambiguous identification [192]. However, the complexity of the real samples data and generation of huge size datasets is always challenging and chemometric tools are used for data compression, resolution, exploration and identification. These multivariate exploratory and resolution methods provide global and local information about chemical components in samples [183, 193]. The advanced data analysis tools provide the possibility of extraction of valuable information from highly complex massive MSI data sets.

DATA PRETREATMENT AND EXPLORATION

Multivariate image analysis (MIA) [194], principal component analysis (PCA) [195], and MCR-ALS [196 - 199] are the most commonly used chemometric methods in the context.

The easy implantation of constraints in MCR-ALS allows reliable extraction of chemical (or biological) information. MCR-ALS application to MSI data is quite simple and interpretation of results provide the distribution maps and mass spectra of the individual resolved components in a simple way. The results give qualitative and semi quantitative relative information about the distribution over different sections of the sample surface [200 - 208].

If case of multiple MS images, extraction of the mixed information within and between samples is more challenging. This situation happens in the case of MSI data of untargeted metabolomics studies that require simultaneous analysis of several images to extract the spatial and structural information on biomarkers. The flowchart of MCR-ALS application for extraction of information from mass spectrometry images of more than one sample is shown in Fig. (**3**). For this purpose, MSI cubic data arrays which have three data modes or directions (x-pixels, y-pixels, and *m/z* values),are unfolded to a data matrix, in a manner that spectra of all *x-y* pixels are put in the rows of matrix on top of each other. In this method, pixel intensities measured at different *m/z* values give a set of column vectors which form a data matrix for each sample. This augmentation is applicable if all samples have the same number of columns, *i.e.*, the number of *m/z* values in all samples are equal. This step is necessary for subsequent bilinear decomposition to obtain distribution maps and mass spectra of analytes [196, 201]. Mass spectra are used for biomarkers (metabolites) identification and distribution maps give their spatial distribution.

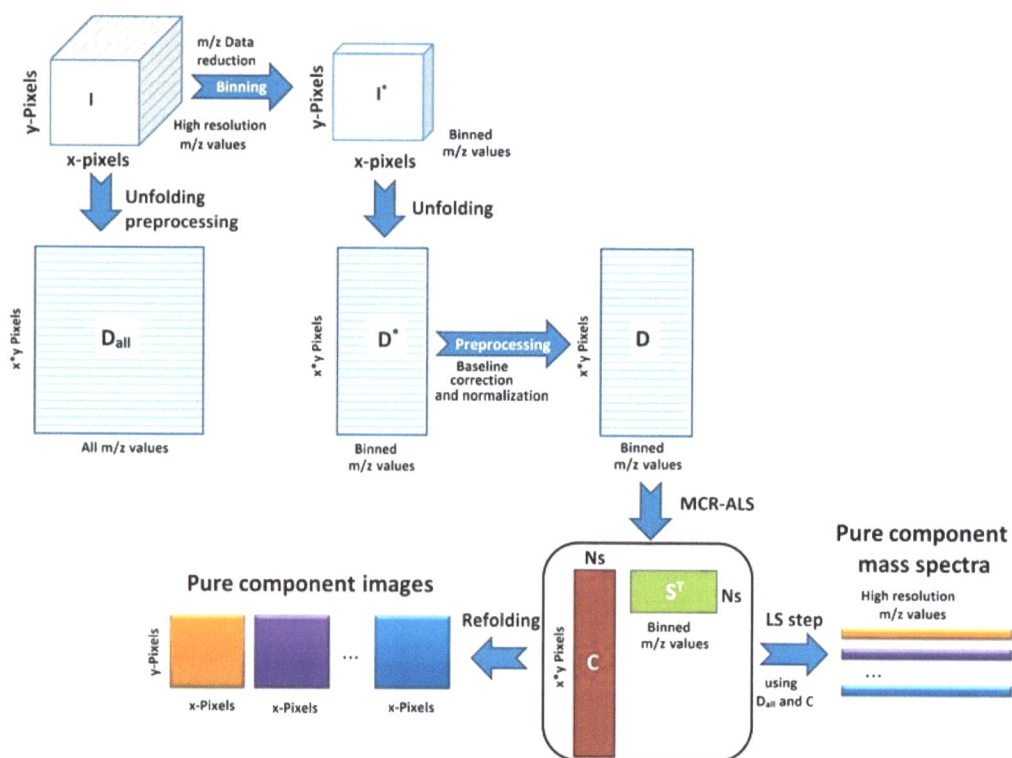

Fig. (3). Flowchart of the simultaneous data pretreatment, data analysis and post processing of more than one sample using MCR-ALS. Reprinted with the permission from Ref. [200].

Fig. (**4**) shows the results obtained from MSI data of a microbe [200]. The applied method in this study is MCR-ALS. As can be seen, MCR-ALS has given the distribution maps of the microbe metabolites and identical mass spectra in all samples in a single step. The distribution maps provide a contour plot type tool to track the spatial presence of the metabolites in different parts of organs or metabolisms. The mass spectra provide a precise tool for identification of metabolites.

DATA COMPRESSION

The vast size of the experimental data sets is truly challenging bottleneck for application of chemometric methods to mass spectrometry imaging (MSI), hence, debottlenecking is necessary. Despite the continuous improvements in calculation power of current laboratory computers, chemometric analysis of MSI data is slow and impractical without use of data compression methods [52, 53]. Among data compression methodologies, the binning and search for regions of interest (ROI) are the most adequate procedures for LC-MS MSI data sets [200].

One of the mostly used procedures for compression of raw LC-MS data is binning. The binning procedure transforms the raw data into a matrix representation, in which retention times are in the *x*-dimension and *m/z* values in the y-dimension. The matrix representation requires the division of the *m/z* axis into equidistant sections having a specific bin size. Each section is analyzed independently. One major drawback of the binning is that binning size is strongly related to the chromatographic profile and proper selection of the bin size for a particular data set is challenging. The small bin size may lead to division of chromatographic peaks between bins and thus detection of analytes becomes difficult. On the contrary, in the case of the large bin size, multiple coelutions between peaks are possible, and small peaks may disappear when the noise is high. Another disadvantage of the binning procedure is the loss of spectral resolution derived from the compression in the *m/z* dimension [13, 209].

Data compression by search for ROI is an satisfactory alternative technique to the binning [210]. Fig. (**5**) shows general scheme of ROI method. This ROI approach is based on the concept of considering analytes as a region of data points having high density. These ROI picks *m/z* values which has intensity higher than a fixed signal-to-noise ratio (S/N) threshold. Also, ROI picks a minimum number of consecutive data points and compresses them within a particular mass deviation, typically set to multiple of the mass accuracy of the mass spectrometer. These constraints prevent ion signals and noise to be considered as an ROI. Using this method, loss of spectral accuracy doesn't occur, which is opposed to spectral accuracy loss in the binning strategy. ROI strategy has been implemented in the *centWave* algorithm of XCMS software [95] and it is increasingly used in feature detection packages as a substitute to the classical binning [209].

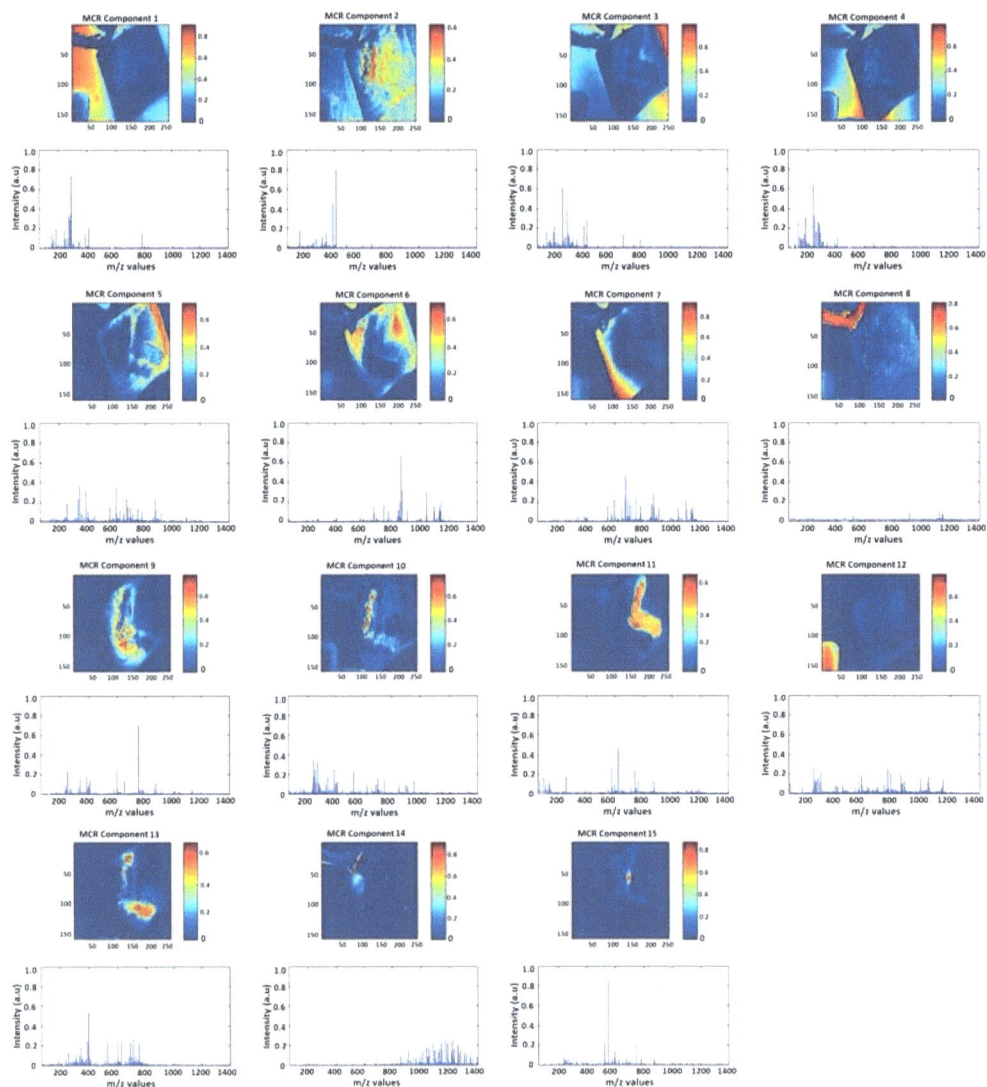

Fig. (4). Identified components of the microbial dataset (distribution maps and corresponding mass Spectra). Reprinted with the permission from Ref [200].

Time windowing is another approach based on the partition of the LC-MS chromatograms into time regions (*i.e.*, time windows) for independent analysis of each window [94, 149, 212]. It is a complementary step to further reduce sample size in case that data compression using binning is not sufficient enough. Usually, the compression level obtained with the ROI strategy is effective enough that analyzing the entire chromatograms in one step is possible.

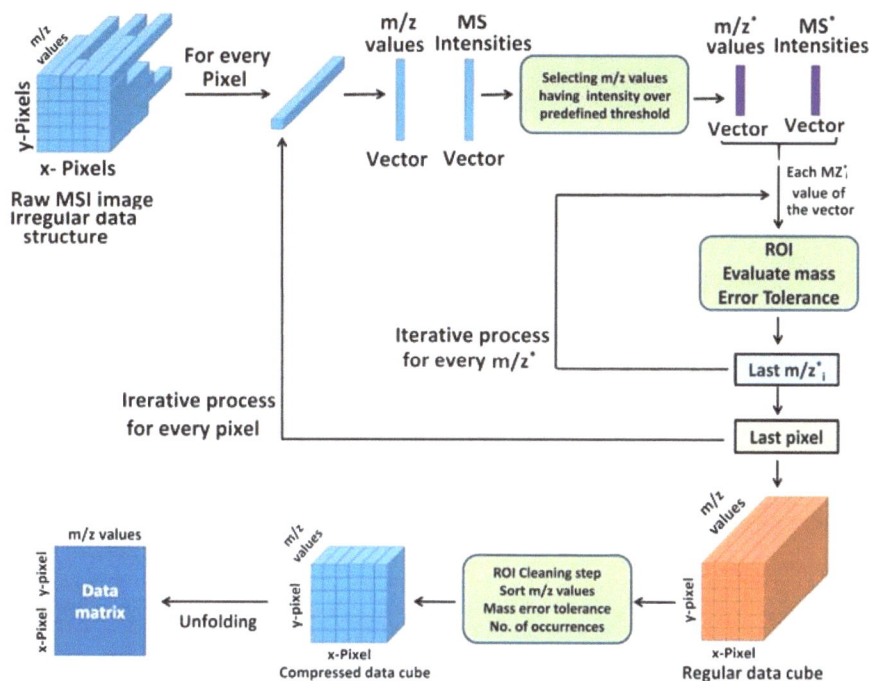

Fig. (5). Flowchart of the algorithm used for mass spectrometry imaging (MSI) data compression based on the detection of the regions of interest (ROIs). Reprinted with permission from Ref [211].

CONCLUDING REMARKS

In this chapter, an overview is given of the application of the chemometrics methodologies for data pretreatment, data mining and data analysis of the mass spectrometry data in life sciences. Because of the technological advances of mass spectrometers and increasing complexity of biological samples, bio (big) data are generated for each sample. In order to get useful information, these data must be analyzed by data analysis strategies (chemometrics). Data pretreatment is used for removing noise and uninformative parts. Then, dependent on the context of the study, the data are modeled to extract useful information. Main fields of application of chemometrics aided mass spectrometry are in omics sciences and biomedical mass imaging. Data in mass based omics are bio (big) and extraction of useful information is almost impossible without chemometrics. Chemometrics presents a complete theory and methodology for every step of omics research, including sampling, experiment design, data pre-processing, metabolite identification, variable selection, and modeling. Application of chemometrics in mass spectrometry imaging is for data compression, data mining and exploration. Data compression is done by binning or ROI procedures, and facilitates the extraction of the information. The compressed data are then mined to find the analytes and their spatial distribution in the under study organism by different chemometrcis methods.

In conclusion, it can be said that the extent of application of the mass spectrometry in life sciences is related to the successful use of chemometrics for these data, otherwise, the vast amount of privileged information in mass data are difficult to extract and hence useless for scientists.

CONSENT FOR PUBLICATION

Not applicable.

CONFLICT OF INTEREST

The authors confirm that this chapter contents have no conflict of interest.

ACKNOWLEDGMENTS

Declared none.

REFERENCES

[1]   Lifshitz, C.; Laskin, J. *Principles of mass spectrometry applied to biomolecules*; John Wiley & Sons: New York, **2006**.

[2]   Hu, Q.; Noll, R.J.; Li, H.; Makarov, A.; Hardman, M.; Graham Cooks, R. The Orbitrap: a new mass spectrometer. *J. Mass Spectrom.,* **2005**, *40*(4), 430-443.
        [http://dx.doi.org/10.1002/jms.856] [PMID: 15838939]

[3]   Horai, H.; Arita, M.; Kanaya, S.; Nihei, Y.; Ikeda, T.; Suwa, K.; Ojima, Y.; Tanaka, K.; Tanaka, S.; Aoshima, K.; Oda, Y.; Kakazu, Y.; Kusano, M.; Tohge, T.; Matsuda, F.; Sawada, Y.; Hirai, M.Y.; Nakanishi, H.; Ikeda, K.; Akimoto, N.; Maoka, T.; Takahashi, H.; Ara, T.; Sakurai, N.; Suzuki, H.; Shibata, D.; Neumann, S.; Iida, T.; Tanaka, K.; Funatsu, K.; Matsuura, F.; Soga, T.; Taguchi, R.; Saito, K.; Nishioka, T. MassBank: a public repository for sharing mass spectral data for life sciences. *J. Mass Spectrom.,* **2010**, *45*(7), 703-714.
        [http://dx.doi.org/10.1002/jms.1777] [PMID: 20623627]

[4]   Jackson, J. *A user's guide to principal components. Wiley, New York, A user's guide to principal components*; John Wiley & Sons: New York, **1991**.
        [http://dx.doi.org/10.1002/0471725331]

[5]   Wold, S.; Ruhe, A.; Wold, H.; Dunn, I. WJ, The collinearity problem in linear regression. The partial least squares (PLS) approach to generalized inverses. *SIAM J. Sci. Comput.,* **1984**, *5*, 735-743.
        [http://dx.doi.org/10.1137/0905052]

[6]   Wold, S.; Sjöström, M.; Eriksson, L. *Partial least squares projections to latent 17 structures (PLS) in chemistry, Encyclopedia of computational chemistry*; , **2002**.
        [http://dx.doi.org/10.1002/0470845015.cpa012]

[7]   Wold, S.; Sjöström, M.; Eriksson, L. PLS-regression: a basic tool of chemometrics. *Chemom. Intell. Lab. Syst.,* **2001**, *58*, 109-130.
        [http://dx.doi.org/10.1016/S0169-7439(01)00155-1]

[8]   Massart, D.L. *Massart, Handbook of chemometrics and qualimetrics, Elsevier, 22 Amesterdam, Netherlands*; , **1997**. 9780080551906

[9]   Otto, M. *Chemometrics: statistics and computer application in analytical chemistry*; John Wiley & Sons, **2016**.
[http://dx.doi.org/10.1002/9783527699377]

[10]   Watson, J.T.; Sparkman, O.D. *Introduction to mass spectrometry: instrumentation, applications, and strategies for data interpretation*; John Wiley & Sons: New York, **2016**.

[11]   Lopes, N.P.; Da Silva, R.R. *Mass Spectrometry in Chemical Biology: Evolving Applications*; Royal Society of Chemistry: London, **2017**.
[http://dx.doi.org/10.1039/9781788010399]

[12]   Varmuza, K. Chemometrics in mass spectrometry. *Int. J. Mass Spectrom. Ion Process.*, **1992**, *118*, 811-823.
[http://dx.doi.org/10.1016/0168-1176(92)85086-F]

[13]   Gorrochategui, E.; Jaumot, J.; Lacorte, S.; Tauler, R. Data analysis strategies for targeted and untargeted LC-MS metabolomic studies: Overview and workflow. *Trends Analyt. Chem.*, **2016**, *82*, 425-442.
[http://dx.doi.org/10.1016/j.trac.2016.07.004]

[14]   Lankadurai, B.P.; Nagato, E.G.; Simpson, M.J. Environmental metabolomics: an emerging approach to study organism responses to environmental stressors. *Environ. Rev.*, **2013**, *21*, 180-205.
[http://dx.doi.org/10.1139/er-2013-0011]

[15]   Navarro-Reig, M.; Jaumot, J.; van Beek, T.A.; Vivó-Truyols, G.; Tauler, R. Chemometric analysis of comprehensive LC×LC-MS data: Resolution of triacylglycerol structural isomers in corn oil. *Talanta*, **2016**, *160*, 624-635.
[http://dx.doi.org/10.1016/j.talanta.2016.08.005] [PMID: 27591659]

[16]   Lu, W.; Bennett, B.D.; Rabinowitz, J.D. Analytical strategies for LC-MS-based targeted metabolomics. *J. Chromatogr. B Analyt. Technol. Biomed. Life Sci.*, **2008**, *871*(2), 236-242.
[http://dx.doi.org/10.1016/j.jchromb.2008.04.031] [PMID: 18502704]

[17]   De Vos, R.C.; Moco, S.; Lommen, A.; Keurentjes, J.J.; Bino, R.J.; Hall, R.D. Untargeted large-scale plant metabolomics using liquid chromatography coupled to mass spectrometry. *Nat. Protoc.*, **2007**, *2*(4), 778-791.
[http://dx.doi.org/10.1038/nprot.2007.95] [PMID: 17446877]

[18]   Gu, H.; Zhang, P.; Zhu, J.; Raftery, D. Globally optimized targeted mass spectrometry: Reliable metabolomics analysis with broad coverage. *Anal. Chem.*, **2015**, *87*(24), 12355-12362.
[http://dx.doi.org/10.1021/acs.analchem.5b03812] [PMID: 26579731]

[19]   Guo, B.; Chen, B.; Liu, A.; Zhu, W.; Yao, S. Liquid chromatography-mass spectrometric multiple reaction monitoring-based strategies for expanding targeted profiling towards quantitative metabolomics. *Curr. Drug Metab.*, **2012**, *13*(9), 1226-1243.
[http://dx.doi.org/10.2174/138920012803341401] [PMID: 22519369]

[20]   Sawada, Y.; Akiyama, K.; Sakata, A.; Kuwahara, A.; Otsuki, H.; Sakurai, T.; Saito, K.; Hirai, M.Y. Widely targeted metabolomics based on large-scale MS/MS data for elucidating metabolite accumulation patterns in plants. *Plant Cell Physiol.*, **2009**, *50*(1), 37-47.
[http://dx.doi.org/10.1093/pcp/pcn183] [PMID: 19054808]

[21]   Dalluge, J.J.; Smith, S.; Sanchez-Riera, F.; McGuire, C.; Hobson, R. Potential of fermentation profiling via rapid measurement of amino acid metabolism by liquid chromatography-tandem mass spectrometry. *J. Chromatogr. A*, **2004**, *1043*(1), 3-7.
[http://dx.doi.org/10.1016/j.chroma.2004.02.010] [PMID: 15317406]

[22]   Wang, S.; Tu, H.; Wan, J.; Chen, W.; Liu, X.; Luo, J.; Xu, J.; Zhang, H. Spatio-temporal distribution and natural variation of metabolites in citrus fruits. *Food Chem.*, **2016**, *199*, 8-17.
[http://dx.doi.org/10.1016/j.foodchem.2015.11.113] [PMID: 26775938]

[23]   Castillo, S.; Gopalacharyulu, P.; Yetukuri, L.; Orešič, M. Algorithms and tools for the preprocessing

of LC–MS metabolomics data. *Chemom. Intell. Lab. Syst.,* **2011**, *108*, 23-32.
[http://dx.doi.org/10.1016/j.chemolab.2011.03.010]

[24] de Raad, M.; Fischer, C.R.; Northen, T.R. High-throughput platforms for metabolomics. *Curr. Opin. Chem. Biol.,* **2016**, *30*, 7-13.
[http://dx.doi.org/10.1016/j.cbpa.2015.10.012] [PMID: 26544850]

[25] Izadmanesh, Y.; Garreta-Lara, E.; Ghasemi, J.B.; Lacorte, S.; Matamoros, V.; Tauler, R. Chemometric analysis of comprehensive two dimensional gas chromatography-mass spectrometry metabolomics data. *J. Chromatogr. A,* **2017**, *1488*, 113-125.
[http://dx.doi.org/10.1016/j.chroma.2017.01.052] [PMID: 28173924]

[26] Ortiz-Villanueva, E.; Jaumot, J.; Benavente, F.; Piña, B.; Sanz-Nebot, V.; Tauler, R. Combination of CE-MS and advanced chemometric methods for high-throughput metabolic profiling. *Electrophoresis,* **2015**, *36*(18), 2324-2335.
[http://dx.doi.org/10.1002/elps.201500027] [PMID: 25820835]

[27] van der Greef, J.; Smilde, A.K. Symbiosis of chemometrics and metabolomics: past, present, and future. *J. Chemometr.,* **2005**, *19*, 376-386.
[http://dx.doi.org/10.1002/cem.941]

[28] Goodacre, R. Making sense of the metabolome using evolutionary computation: seeing the wood with the trees. *J. Exp. Bot.,* **2005**, *56*(410), 245-254.
[http://dx.doi.org/10.1093/jxb/eri043] [PMID: 15596480]

[29] Lundstedt, T.; Seifert, E.; Abramo, L.; Thelin, B.; Nyström, Å.; Pettersen, J.; Bergman, R. Experimental design and optimization. *Chemom. Intell. Lab. Syst.,* **1998**, *42*, 3-40.
[http://dx.doi.org/10.1016/S0169-7439(98)00065-3]

[30] Box, G.E.; Hunter, W.G.; Hunter, J.S. *Statistics for experimenters*; John wiley & Sons: New York, **1978**.

[31] Johansson, E.; Kettaneh-Wold, N.; Wikstrom, C.; Wold, S.; Ericksson, L. *Design of Experiments, Principles and Applications, Umetrics Academy, Umeå,*; , **2000**. ISBN: ISBN-13: 978-9197373005

[32] Dunn, W.B.; Wilson, I.D.; Nicholls, A.W.; Broadhurst, D. The importance of experimental design and QC samples in large-scale and MS-driven untargeted metabolomic studies of humans. *Bioanalysis,* **2012**, *4*(18), 2249-2264.
[http://dx.doi.org/10.4155/bio.12.204] [PMID: 23046267]

[33] León, Z.; García-Cañaveras, J.C.; Donato, M.T.; Lahoz, A. Mammalian cell metabolomics: experimental design and sample preparation. *Electrophoresis,* **2013**, *34*(19), 2762-2775.
[http://dx.doi.org/10.1002/elps.201200605] [PMID: 23436493]

[34] Smith, C.A.; Want, E.J.; O'Maille, G.; Abagyan, R.; Siuzdak, G. XCMS: processing mass spectrometry data for metabolite profiling using nonlinear peak alignment, matching, and identification. *Anal. Chem.,* **2006**, *78*(3), 779-787.
[http://dx.doi.org/10.1021/ac051437y] [PMID: 16448051]

[35] Benton, H.P.; Wong, D.M.; Trauger, S.A.; Siuzdak, G. XCMS2: processing tandem mass spectrometry data for metabolite identification and structural characterization. *Anal. Chem.,* **2008**, *80*(16), 6382-6389.
[http://dx.doi.org/10.1021/ac800795f] [PMID: 18627180]

[36] Katajamaa, M.; Miettinen, J.; Orešič, M. MZmine: toolbox for processing and visualization of mass spectrometry based molecular profile data. *Bioinformatics,* **2006**, *22*(5), 634-636.
[http://dx.doi.org/10.1093/bioinformatics/btk039] [PMID: 16403790]

[37] Pluskal, T.; Castillo, S.; Villar-Briones, A.; Orešič, M. MZmine 2: modular framework for processing, visualizing, and analyzing mass spectrometry-based molecular profile data. *BMC Bioinformatics,* **2010**, *11*, 395.
[http://dx.doi.org/10.1186/1471-2105-11-395] [PMID: 20650010]

[38]   Allwood, J.W.; Goodacre, R. An introduction to liquid chromatography-mass spectrometry instrumentation applied in plant metabolomic analyses. *Phytochem. Anal.,* **2010**, *21*(1), 33-47.
[http://dx.doi.org/10.1002/pca.1187] [PMID: 19927296]

[39]   De Vos, R.C.H.; Moco, S.; Lommen, A.; Keurentjes, J.J.B.; Bino, R.J.; Hall, R.D. Untargeted large-scale plant metabolomics using liquid chromatography coupled to mass spectrometry. *Nat. Protoc.,* **2007**, *2*(4), 778-791.
[http://dx.doi.org/10.1038/nprot.2007.95] [PMID: 17446877]

[40]   Wei, X.; Sun, W.; Shi, X.; Koo, I.; Wang, B.; Zhang, J.; Yin, X.; Tang, Y.; Bogdanov, B.; Kim, S.; Zhou, Z.; McClain, C.; Zhang, X. MetSign: a computational platform for high-resolution mass spectrometry-based metabolomics. *Anal. Chem.,* **2011**, *83*(20), 7668-7675.
[http://dx.doi.org/10.1021/ac2017025] [PMID: 21932828]

[41]   Duran, A.L.; Yang, J.; Wang, L.; Sumner, L.W. Metabolomics spectral formatting, alignment and conversion tools (MSFACTs). *Bioinformatics,* **2003**, *19*(17), 2283-2293.
[http://dx.doi.org/10.1093/bioinformatics/btg315] [PMID: 14630657]

[42]   Katajamaa, M.; Orešič, M. Data processing for mass spectrometry-based metabolomics. *J. Chromatogr. A,* **2007**, *1158*(1-2), 318-328.
[http://dx.doi.org/10.1016/j.chroma.2007.04.021] [PMID: 17466315]

[43]   Zhang, Z-M.; Chen, S.; Liang, Y-Z. Baseline correction using adaptive iteratively reweighted penalized least squares. *Analyst (Lond.),* **2010**, *135*(5), 1138-1146.
[http://dx.doi.org/10.1039/b922045c] [PMID: 20419267]

[44]   Yi, L.; Dong, N.; Yun, Y.; Deng, B.; Ren, D.; Liu, S.; Liang, Y. Chemometric methods in data processing of mass spectrometry-based metabolomics: A review. *Anal. Chim. Acta,* **2016**, *914*, 17-34.
[http://dx.doi.org/10.1016/j.aca.2016.02.001] [PMID: 26965324]

[45]   Hilario, M.; Kalousis, A.; Pellegrini, C.; Müller, M. Processing and classification of protein mass spectra. *Mass Spectrom. Rev.,* **2006**, *25*(3), 409-449.
[http://dx.doi.org/10.1002/mas.20072] [PMID: 16463283]

[46]   Haimi, P.; Uphoff, A.; Hermansson, M.; Somerharju, P. Software tools for analysis of mass spectrometric lipidome data. *Anal. Chem.,* **2006**, *78*(24), 8324-8331.
[http://dx.doi.org/10.1021/ac061390w] [PMID: 17165823]

[47]   Bellew, M.; Coram, M.; Fitzgibbon, M.; Igra, M.; Randolph, T.; Wang, P.; May, D.; Eng, J.; Fang, R.; Lin, C.; Chen, J.; Goodlett, D.; Whiteaker, J.; Paulovich, A.; McIntosh, M. A suite of algorithms for the comprehensive analysis of complex protein mixtures using high-resolution LC-MS. *Bioinformatics,* **2006**, *22*(15), 1902-1909.
[http://dx.doi.org/10.1093/bioinformatics/btl276] [PMID: 16766559]

[48]   Kiers, H.A.L.; ten Berge, J.M.F.; Bro, R. PARAFAC2—Part I. A direct fitting algorithm for the PARAFAC2 model. *J. Chemometr.,* **1999**, *13*, 275-294.
[http://dx.doi.org/10.1002/(SICI)1099-128X(199905/08)13:3/4<275::AID-CEM543>3.0.CO;2-B]

[49]   Navarro-Reig, M.; Jaumot, J.; Tauler, R. An untargeted lipidomic strategy combining comprehensive two-dimensional liquid chromatography and chemometric analysis. *J. Chromatogr. A,* **2018**, *1568*, 80-90.
[http://dx.doi.org/10.1016/j.chroma.2018.07.017] [PMID: 30001900]

[50]   Gorrochategui, E.; Jaumot, J.; Tauler, R. A protocol for LC-MS metabolomic data processing using chemometric tools. *Protoc. Exch.,* **2015**.
[http://dx.doi.org/10.1038/protex.2015.102]

[51]   Jaumot, J.; de Juan, A.; Tauler, R. MCR-ALS GUI 2.0: new features and applications. *Chemom. Intell. Lab. Syst.,* **2015**, *140*, 1-12.
[http://dx.doi.org/10.1016/j.chemolab.2014.10.003]

[52]   Navarro-Reig, M.; Bedia, C.; Tauler, R.; Jaumot, J. Chemometric Strategies for Peak Detection and

Profiling from Multidimensional Chromatography. *Proteomics,* **2018**, *18*(18)e1700327
[http://dx.doi.org/10.1002/pmic.201700327] [PMID: 29611629]

[53]   Navarro-Reig, M.; Jaumot, J.; Baglai, A.; Vivó-Truyols, G.; Schoenmakers, P.J.; Tauler, R. Untargeted comprehensive two-dimensional liquid chromatography coupled with high-resolution mass spectrometry analysis of rice metabolome using multivariate curve resolution. *Anal. Chem.,* **2017**, *89*(14), 7675-7683.
[http://dx.doi.org/10.1021/acs.analchem.7b01648] [PMID: 28643516]

[54]   Wang, W.; Zhou, H.; Lin, H.; Roy, S.; Shaler, T.A.; Hill, L.R.; Norton, S.; Kumar, P.; Anderle, M.; Becker, C.H. Quantification of proteins and metabolites by mass spectrometry without isotopic labeling or spiked standards. *Anal. Chem.,* **2003**, *75*(18), 4818-4826.
[http://dx.doi.org/10.1021/ac026468x] [PMID: 14674459]

[55]   van den Berg, R.A.; Hoefsloot, H.C.; Westerhuis, J.A.; Smilde, A.K.; van der Werf, M.J. Centering, scaling, and transformations: improving the biological information content of metabolomics data. *BMC Genomics,* **2006**, *7*, 142.
[http://dx.doi.org/10.1186/1471-2164-7-142] [PMID: 16762068]

[56]   Kvalheim, O.M.; Brakstad, F.; Liang, Y. Preprocessing of analytical profiles in the presence of homoscedastic or heteroscedastic noise. *Anal. Chem.,* **1994**, *66*, 43-51.
[http://dx.doi.org/10.1021/ac00073a010]

[57]   Watson, D.G. A rough guide to metabolite identification using high resolution liquid chromatography mass spectrometry in metabolomic profiling in metazoans. *Comput. Struct. Biotechnol. J.,* **2013**, 4e201301005
[http://dx.doi.org/10.5936/csbj.201301005] [PMID: 24688687]

[58]   Holčapek, M.; Jirásko, R.; Lísa, M. Basic rules for the interpretation of atmospheric pressure ionization mass spectra of small molecules. *J. Chromatogr. A,* **2010**, *1217*(25), 3908-3921.
[http://dx.doi.org/10.1016/j.chroma.2010.02.049] [PMID: 20303090]

[59]   Wishart, D.S. Computational strategies for metabolite identification in metabolomics. *Bioanalysis,* **2009**, *1*(9), 1579-1596.
[http://dx.doi.org/10.4155/bio.09.138] [PMID: 21083105]

[60]   Kind, T.; Fiehn, O. Advances in structure elucidation of small molecules using mass spectrometry. *Bioanal. Rev.,* **2010**, *2*(1-4), 23-60.
[http://dx.doi.org/10.1007/s12566-010-0015-9] [PMID: 21289855]

[61]   Stein, S.E.; Scott, D.R. Optimization and testing of mass spectral library search algorithms for compound identification. *J. Am. Soc. Mass Spectrom.,* **1994**, *5*(9), 859-866.
[http://dx.doi.org/10.1016/1044-0305(94)87009-8] [PMID: 24222034]

[62]   Koo, I.; Kim, S.; Zhang, X. Comparative analysis of mass spectral matching-based compound identification in gas chromatography-mass spectrometry. *J. Chromatogr. A,* **2013**, *1298*, 132-138.
[http://dx.doi.org/10.1016/j.chroma.2013.05.021] [PMID: 23726352]

[63]   Dunn, W.B.; Broadhurst, D.; Begley, P.; Zelena, E.; Francis-McIntyre, S.; Anderson, N.; Brown, M.; Knowles, J.D.; Halsall, A.; Haselden, J.N.; Nicholls, A.W.; Wilson, I.D.; Kell, D.B.; Goodacre, R. Procedures for large-scale metabolic profiling of serum and plasma using gas chromatography and liquid chromatography coupled to mass spectrometry. *Nat. Protoc.,* **2011**, *6*(7), 1060-1083.
[http://dx.doi.org/10.1038/nprot.2011.335] [PMID: 21720319]

[64]   Kopka, J. Current challenges and developments in GC-MS based metabolite profiling technology. *J. Biotechnol.,* **2006**, *124*(1), 312-322.
[http://dx.doi.org/10.1016/j.jbiotec.2005.12.012] [PMID: 16434119]

[65]   Benecke, C.; Grund, R.; Hohberger, R.; Kerber, A.; Laue, R.; Wieland, T. MOLGEN+, a generator of connectivity isomers and stereoisomers for molecular structure elucidation. *Anal. Chim. Acta,* **1995**, *314*, 141-147.
[http://dx.doi.org/10.1016/0003-2670(95)00291-7]

[66]   Peironcely, J.E.; Rojas-Chertó, M.; Fichera, D.; Reijmers, T.; Coulier, L.; Faulon, J-L.; Hankemeier, T. OMG: Open Molecule Generator. *J. Cheminform.*, **2012**, *4*(1), 21.
        [http://dx.doi.org/10.1186/1758-2946-4-21] [PMID: 22985496]

[67]   Stein, S.E. Chemical substructure identification by mass spectral library searching. *J. Am. Soc. Mass Spectrom.*, **1995**, *6*(8), 644-655.
        [http://dx.doi.org/10.1016/1044-0305(95)00291-K] [PMID: 24214391]

[68]   Halket, J.M.; Waterman, D.; Przyborowska, A.M.; Patel, R.K.P.; Fraser, P.D.; Bramley, P.M. Chemical derivatization and mass spectral libraries in metabolic profiling by GC/MS and LC/MS/MS. *J. Exp. Bot.*, **2005**, *56*(410), 219-243.
        [http://dx.doi.org/10.1093/jxb/eri069] [PMID: 15618298]

[69]   Kind, T.; Fiehn, O. Seven Golden Rules for heuristic filtering of molecular formulas obtained by accurate mass spectrometry. *BMC Bioinformatics*, **2007**, *8*, 105.
        [http://dx.doi.org/10.1186/1471-2105-8-105] [PMID: 17389044]

[70]   Erve, J.C.L.; Gu, M.; Wang, Y.; DeMaio, W.; Talaat, R.E. Spectral accuracy of molecular ions in an LTQ/Orbitrap mass spectrometer and implications for elemental composition determination. *J. Am. Soc. Mass Spectrom.*, **2009**, *20*(11), 2058-2069.
        [http://dx.doi.org/10.1016/j.jasms.2009.07.014] [PMID: 19716315]

[71]   Wang, Y.; Gu, M. The concept of spectral accuracy for MS. *Anal. Chem.*, **2010**, *82*(17), 7055-7062.
        [http://dx.doi.org/10.1021/ac100888b] [PMID: 20684651]

[72]   Nagao, T.; Yukihira, D.; Fujimura, Y.; Saito, K.; Takahashi, K.; Miura, D.; Wariishi, H. Power of isotopic fine structure for unambiguous determination of metabolite elemental compositions: in silico evaluation and metabolomic application. *Anal. Chim. Acta*, **2014**, *813*, 70-76.
        [http://dx.doi.org/10.1016/j.aca.2014.01.032] [PMID: 24528662]

[73]   Zhu, Z-J.; Schultz, A.W.; Wang, J.; Johnson, C.H.; Yannone, S.M.; Patti, G.J.; Siuzdak, G. Liquid chromatography quadrupole time-of-flight mass spectrometry characterization of metabolites guided by the METLIN database. *Nat. Protoc.*, **2013**, *8*(3), 451-460.
        [http://dx.doi.org/10.1038/nprot.2013.004] [PMID: 23391889]

[74]   Little, J.L.; Williams, A.J.; Pshenichnov, A.; Tkachenko, V. Identification of "known unknowns" utilizing accurate mass data and ChemSpider. *J. Am. Soc. Mass Spectrom.*, **2012**, *23*(1), 179-185.
        [http://dx.doi.org/10.1007/s13361-011-0265-y] [PMID: 22069037]

[75]   Stein, S. Mass spectral reference libraries: an ever-expanding resource for chemical identification. *Anal. Chem.*, **2012**, *84*(17), 7274-7282.
        [http://dx.doi.org/10.1021/ac301205z] [PMID: 22803687]

[76]   Werner, E.; Heilier, J-F.; Ducruix, C.; Ezan, E.; Junot, C.; Tabet, J-C. Mass spectrometry for the identification of the discriminating signals from metabolomics: current status and future trends. *J. Chromatogr. B Analyt. Technol. Biomed. Life Sci.*, **2008**, *871*(2), 143-163.
        [http://dx.doi.org/10.1016/j.jchromb.2008.07.004] [PMID: 18672410]

[77]   Hufsky, F.; Scheubert, K.; Böcker, S. Computational mass spectrometry for small-molecule fragmentation. *Trends Analyt. Chem.*, **2014**, *53*, 41-48.
        [http://dx.doi.org/10.1016/j.trac.2013.09.008]

[78]   Huan, T.; Tang, C.; Li, R.; Shi, Y.; Lin, G.; Li, L. MyCompoundID MS/MS Search: Metabolite Identification Using a Library of Predicted Fragment-Ion-Spectra of 383,830 Possible Human Metabolites. *Anal. Chem.*, **2015**, *87*(20), 10619-10626.
        [http://dx.doi.org/10.1021/acs.analchem.5b03126] [PMID: 26415007]

[79]   Kangas, L.J.; Metz, T.O.; Isaac, G.; Schrom, B.T.; Ginovska-Pangovska, B.; Wang, L.; Tan, L.; Lewis, R.R.; Miller, J.H. In silico identification software (ISIS): a machine learning approach to tandem mass spectral identification of lipids. *Bioinformatics*, **2012**, *28*(13), 1705-1713.
        [http://dx.doi.org/10.1093/bioinformatics/bts194] [PMID: 22592377]

[80]   Hill, A.W.; Mortishire-Smith, R.J. Automated assignment of high-resolution collisionally activated dissociation mass spectra using a systematic bond disconnection approach. *Rapid Commun. Mass Spectrom.,* **2005**, *19*, 3111-3118.
[http://dx.doi.org/10.1002/rcm.2177]

[81]   Bonn, B.; Leandersson, C.; Fontaine, F.; Zamora, I. Enhanced metabolite identification with MS(E) and a semi-automated software for structural elucidation. *Rapid Commun. Mass Spectrom.,* **2010**, *24*(21), 3127-3138.
[http://dx.doi.org/10.1002/rcm.4753] [PMID: 20941759]

[82]   Wolf, S.; Schmidt, S.; Müller-Hannemann, M.; Neumann, S. In silico fragmentation for computer assisted identification of metabolite mass spectra. *BMC Bioinformatics,* **2010**, *11*, 148.
[http://dx.doi.org/10.1186/1471-2105-11-148] [PMID: 20307295]

[83]   Heinonen, M.; Shen, H.; Zamboni, N.; Rousu, J. Metabolite identification and molecular fingerprint prediction through machine learning. *Bioinformatics,* **2012**, *28*(18), 2333-2341.
[http://dx.doi.org/10.1093/bioinformatics/bts437] [PMID: 22815355]

[84]   Allen, F.; Greiner, R.; Wishart, D. Competitive fragmentation modeling of ESI-MS/MS spectra for putative metabolite identification. *Metabolomics,* **2015**, *11*, 98-110.
[http://dx.doi.org/10.1007/s11306-014-0676-4]

[85]   Böcker, S.; Rasche, F. Towards de novo identification of metabolites by analyzing tandem mass spectra. *Bioinformatics,* **2008**, *24*(16), i49-i55.
[http://dx.doi.org/10.1093/bioinformatics/btn270] [PMID: 18689839]

[86]   Rasche, F.; Svatoš, A.; Maddula, R.K.; Böttcher, C.; Böcker, S. Computing fragmentation trees from tandem mass spectrometry data. *Anal. Chem.,* **2011**, *83*(4), 1243-1251.
[http://dx.doi.org/10.1021/ac101825k] [PMID: 21182243]

[87]   Hufsky, F.; Rempt, M.; Rasche, F.; Pohnert, G.; Böcker, S. De novo analysis of electron impact mass spectra using fragmentation trees. *Anal. Chim. Acta,* **2012**, *739*, 67-76.
[http://dx.doi.org/10.1016/j.aca.2012.06.021] [PMID: 22819051]

[88]   Rauf, I.; Rasche, F.; Nicolas, F.; Böcker, S. *Böcker, Finding Maximum Colorful 31 Subtrees in Practice, Springer Berlin Heidelberg, Berlin, Heidelberg*; , **2012**, p. 213-223.

[89]   Ridder, L.; van der Hooft, J.J.; Verhoeven, S.; de Vos, R.C.; van Schaik, R.; Vervoort, J. Substructure-based annotation of high-resolution multistage MS(n) spectral trees. *Rapid Commun. Mass Spectrom.,* **2012**, *26*(20), 2461-2471.
[http://dx.doi.org/10.1002/rcm.6364] [PMID: 22976213]

[90]   Johnson, C.H.; Ivanisevic, J.; Benton, H.P.; Siuzdak, G. Bioinformatics: the next frontier of metabolomics. *Anal. Chem.,* **2015**, *87*(1), 147-156.
[http://dx.doi.org/10.1021/ac5040693] [PMID: 25389922]

[91]   Tautenhahn, R.; Patti, G.J.; Rinehart, D.; Siuzdak, G. XCMS Online: a web-based platform to process untargeted metabolomic data. *Anal. Chem.,* **2012**, *84*(11), 5035-5039.
[http://dx.doi.org/10.1021/ac300698c] [PMID: 22533540]

[92]   Katajamaa, M.; Miettinen, J.; Oresic, M. MZmine: toolbox for processing and visualization of mass spectrometry based molecular profile data. *Bioinformatics,* **2006**, *22*(5), 634-636.
[http://dx.doi.org/10.1093/bioinformatics/btk039] [PMID: 16403790]

[93]   Pluskal, T.; Castillo, S.; Villar-Briones, A.; Oresic, M. MZmine 2: modular framework for processing, visualizing, and analyzing mass spectrometry-based molecular profile data. *BMC Bioinformatics,* **2010**, *11*, 395.
[http://dx.doi.org/10.1186/1471-2105-11-395] [PMID: 20650010]

[94]   Farrés, M.; Piña, B.; Tauler, R. Chemometric evaluation of *Saccharomyces cerevisiae* metabolic profiles using LC-MS. *Metabolomics,* **2015**, *11*, 210-224.
[http://dx.doi.org/10.1007/s11306-014-0689-z] [PMID: 25598766]

[95] Smith, C.A.; Want, E.J.; O'Maille, G.; Abagyan, R.; Siuzdak, G. XCMS: processing mass spectrometry data for metabolite profiling using nonlinear peak alignment, matching, and identification. *Anal. Chem.,* **2006**, *78*(3), 779-787.
[http://dx.doi.org/10.1021/ac051437y] [PMID: 16448051]

[96] Keller, A. J. Eng, N. Zhang, X.J. Li, R. Aebersold, A uniform proteomics 13 MS/MS analysis platform utilizing open XML file formats *Mol. Syst. Biol., 1,* **2005**, 2005.0017.

[97] Gehlenborg, N.; Yan, W.; Lee, I.Y.; Yoo, H.; Nieselt, K.; Hwang, D.; Aebersold, R.; Hood, L. Prequips--an extensible software platform for integration, visualization and analysis of LC-MS/MS proteomics data. *Bioinformatics,* **2009**, *25*(5), 682-683.
[http://dx.doi.org/10.1093/bioinformatics/btp005] [PMID: 19129212]

[98] Kohlbacher, O.; Reinert, K.; Gröpl, C.; Lange, E.; Pfeifer, N.; Schulz-Trieglaff, O.; Sturm, M. TOPP--the OpenMS proteomics pipeline. *Bioinformatics,* **2007**, *23*(2), e191-e197.
[http://dx.doi.org/10.1093/bioinformatics/btl299] [PMID: 17237091]

[99] Kessner, D.; Chambers, M.; Burke, R.; Agus, D.; Mallick, P. ProteoWizard: open source software for rapid proteomics tools development. *Bioinformatics,* **2008**, *24*(21), 2534-2536.
[http://dx.doi.org/10.1093/bioinformatics/btn323] [PMID: 18606607]

[100] Boccard, J.; Veuthey, J-L.; Rudaz, S. Knowledge discovery in metabolomics: an overview of MS data handling. *J. Sep. Sci.,* **2010**, *33*(3), 290-304.
[http://dx.doi.org/10.1002/jssc.200900609] [PMID: 20087872]

[101] Methods for Variable Ranking and Selection, John Wiley & Sons, New 28 York.
[http://dx.doi.org/10.1002/9783527677320.]

[102] Mehmood, T.; Liland, K.H.; Snipen, L.; Sæbø, S. A review of variable selection methods in Partial Least Squares Regression. *Chemom. Intell. Lab. Syst.,* **2012**, *118*, 62-69.
[http://dx.doi.org/10.1016/j.chemolab.2012.07.010]

[103] Favilla, S.; Durante, C.; Vigni, M.L.; Cocchi, M. Assessing feature relevance in NPLS models by VIP. *Chemom. Intell. Lab. Syst.,* **2013**, *129*, 76-86.
[http://dx.doi.org/10.1016/j.chemolab.2013.05.013]

[104] Rajalahti, T.; Arneberg, R.; Kroksveen, A.C.; Berle, M.; Myhr, K-M.; Kvalheim, O.M. Discriminating variable test and selectivity ratio plot: quantitative tools for interpretation and variable (biomarker) selection in complex spectral or chromatographic profiles. *Anal. Chem.,* **2009**, *81*(7), 2581-2590.
[http://dx.doi.org/10.1021/ac802514y] [PMID: 19228047]

[105] Kvalheim, O.M. Interpretation of partial least squares regression models by means of target projection and selectivity ratio plots. *J. Chemometr.,* **2010**, *24*, 496-504.
[http://dx.doi.org/10.1002/cem.1289]

[106] Yi, L.; Dong, N.; Shi, S.; Deng, B.; Yun, Y.; Yi, Z.; Zhang, Y. Metabolomic identification of novel biomarkers of nasopharyngeal carcinoma. *RSC Advances,* **2014**, *4*, 59094-59101.
[http://dx.doi.org/10.1039/C4RA09860A]

[107] Yun, Y-H.; Liang, F.; Deng, B-C.; Lai, G-B.; Vicente Gonçalves, C.M.; Lu, H-M.; Yan, J.; Huang, X.; Yi, L-Z.; Liang, Y-Z. Informative metabolites identification by variable importance analysis based on random variable combination. *Metabolomics,* **2015**, *11*, 1539-1551.
[http://dx.doi.org/10.1007/s11306-015-0803-x]

[108] Farrés, M.; Platikanov, S.; Tsakovski, S.; Tauler, R. Comparison of the variable importance in projection (VIP) and of the selectivity ratio (SR) methods for variable selection and interpretation. *J. Chemometr.,* **2015**, *29*, 528-536.
[http://dx.doi.org/10.1002/cem.2736]

[109] Yun, Y-H.; Deng, B-C.; Cao, D-S.; Wang, W-T.; Liang, Y-Z. Variable importance analysis based on rank aggregation with applications in metabolomics for biomarker discovery. *Anal. Chim. Acta,* **2016**, *911*, 27-34.

[http://dx.doi.org/10.1016/j.aca.2015.12.043] [PMID: 26893083]

[110] Guyon, I.; Elisseeff, A. An introduction to variable and feature selection. *J. Mach. Learn. Res.,* **2003**, *3*, 1157-1182.

[111] Correa, E.; Goodacre, R. A genetic algorithm-Bayesian network approach for the analysis of metabolomics and spectroscopic data: application to the rapid identification of Bacillus spores and classification of Bacillus species. *BMC Bioinformatics,* **2011**, *12*, 33.
[http://dx.doi.org/10.1186/1471-2105-12-33] [PMID: 21269434]

[112] Anastassiou, D. Computational analysis of the synergy among multiple interacting genes. *Mol. Syst. Biol.,* **2007**, *3*, 83.
[http://dx.doi.org/10.1038/msb4100124] [PMID: 17299419]

[113] Breiman, L. Random Forests. *Mach. Learn.,* **2001**, *45*, 5-32.
[http://dx.doi.org/10.1023/A:1010933404324]

[114] Li, H-D.; Liang, Y-Z.; Xu, Q-S.; Cao, D-S. Model population analysis for variable selection. *J. Chemometr.,* **2010**, *24*, 418-423.
[http://dx.doi.org/10.1002/cem.1300]

[115] Li, H-D.; Zeng, M-M.; Tan, B-B.; Liang, Y-Z.; Xu, Q-S.; Cao, D-S. Recipe for revealing informative metabolites based on model population analysis. *Metabolomics,* **2010**, *6*, 353-361.
[http://dx.doi.org/10.1007/s11306-010-0213-z]

[116] Li, H.D.; Liang, Y.Z.; Xu, Q.S.; Cao, D.S.; Tan, B.B.; Deng, B.C.; Lin, C.C. Recipe for uncovering predictive genes using support vector machines based on model population analysis. *IEEE/ACM Trans. Comput. Biol. Bioinformatics,* **2011**, *8*(6), 1633-1641.
[http://dx.doi.org/10.1109/TCBB.2011.36] [PMID: 21339535]

[117] Zhang, H.; Li, H.; Boyles, M.J.; Henschel, R.; Kohara, E.K.; Ando, M. Exploiting HPC resources for the 3D-time series analysis of caries lesion activity *Proceedings of the 1st Conference of the Extreme Science and Engineering Discovery Environment: Bridging from the eXtreme to the campus and beyond,* **2012**, , pp. 1-8.
[http://dx.doi.org/10.1145/2335755.2335815]

[118] Deng, B.C.; Yun, Y.H.; Liang, Y.Z.; Yi, L.Z. A novel variable selection approach that iteratively optimizes variable space using weighted binary matrix sampling. *Analyst (Lond.),* **2014**, *139*(19), 4836-4845.
[http://dx.doi.org/10.1039/C4AN00730A] [PMID: 25083512]

[119] Li, H-D.; Xu, Q-S.; Zhang, W.; Liang, Y-Z. Variable complementary network: a novel approach for identifying biomarkers and their mutual associations. *Metabolomics,* **2012**, *8*, 1218-1226.
[http://dx.doi.org/10.1007/s11306-012-0410-z]

[120] Hall, M.A. Correlation-based feature selection for machine learning, PhD 8 Thesis, The University of Waikato, Hamilton, New Ziland **1999**.

[121] Ben-Bassat, M. *Pattern recognition and reduction of dimensionality, 10 Handbook of Statistics, Elsevier, Amesterdam, Netherlands*; , **1982**, 2, p. 773-910..

[122] Liang, J.; Yang, S.; Winstanley, A. Invariant optimal feature selection: A distance discriminant and feature ranking based solution. *Pattern Recognit.,* **2008**, *41*, 1429-1439.
[http://dx.doi.org/10.1016/j.patcog.2007.10.018]

[123] Yu, L.; Liu, H. Efficient feature selection via analysis of relevance and redundancy. *J. Mach. Learn. Res.,* **2004**, *5*, 1205-1224.

[124] Li, H.; Liang, Y.; Xu, Q.; Cao, D. Key wavelengths screening using competitive adaptive reweighted sampling method for multivariate calibration. *Anal. Chim. Acta,* **2009**, *648*(1), 77-84.
[http://dx.doi.org/10.1016/j.aca.2009.06.046] [PMID: 19616692]

[125] Cao, M.D.; Sitter, B.; Bathen, T.F.; Bofin, A.; Lønning, P.E.; Lundgren, S.; Gribbestad, I.S. Predicting

long-term survival and treatment response in breast cancer patients receiving neoadjuvant chemotherapy by MR metabolic profiling. *NMR Biomed.,* **2012,** *25*(2), 369-378.
[http://dx.doi.org/10.1002/nbm.1762] [PMID: 21823183]

[126] Alba, E.; Garcia-Nieto, J.; Jourdan, L.; Talbi, E. Gene selection in cancer classification using PSO/SVM and GA/SVM hybrid algorithms *IEEE Congress on Evolutionary Computation,* **2007,** pp. 284-290.
[http://dx.doi.org/10.1109/CEC.2007.4424483]

[127] Zhao, Z.; Liu, H. Searching for interacting features in subset selection. *Intell. Data Anal.,* **2009,** *13*, 207-228.
[http://dx.doi.org/10.3233/IDA-2009-0364]

[128] Mao, Q.; Bai, M.; Xu, J-D.; Kong, M.; Zhu, L-Y.; Zhu, H.; Wang, Q.; Li, S-L. Discrimination of leaves of Panax ginseng and P. quinquefolius by ultra high performance liquid chromatography quadrupole/time-of-flight mass spectrometry based metabolomics approach. *J. Pharm. Biomed. Anal.,* **2014,** *97*, 129-140.
[http://dx.doi.org/10.1016/j.jpba.2014.04.032] [PMID: 24867296]

[129] Wang, J.; Reijmers, T.; Chen, L.; Van Der Heijden, R.; Wang, M.; Peng, S.; Hankemeier, T.; Xu, G.; Van Der Greef, J. Systems toxicology study of doxorubicin on rats using ultra performance liquid chromatography coupled with mass spectrometry based metabolomics. *Metabolomics,* **2009,** *5*(4), 407-418.
[http://dx.doi.org/10.1007/s11306-009-0165-3] [PMID: 20046867]

[130] Draisma, H.H.M.; Reijmers, T.H.; Meulman, J.J.; van der Greef, J.; Hankemeier, T.; Boomsma, D.I. Hierarchical clustering analysis of blood plasma lipidomics profiles from mono- and dizygotic twin families. *Eur. J. Hum. Genet.,* **2013,** *21*(1), 95-101.
[http://dx.doi.org/10.1038/ejhg.2012.110] [PMID: 22713803]

[131] Kriegel, H-P.; Kröger, P.; Zimek, A. Clustering high-dimensional data: A survey on subspace clustering, pattern-based clustering, and correlation clustering. *ACM Trans. Knowl. Discov. Data,* **2009,** *3*, 1.
[http://dx.doi.org/10.1145/1497577.1497578]

[132] Cody, R. Goodwin, Brett C. Covington, Dagmara K. Derewacz, C.R. McNees, John P. Wikswo, John A. McLean, Brian O. Bachmann, Structuring Microbial Metabolic Responses to Multiplexed Stimuli via Self-Organizing Metabolomics Maps. *Chem. Biol.,* **2015,** *22*, 661-670.
[http://dx.doi.org/10.1016/j.chembiol.2015.03.020]

[133] Kim, J.W.; Monila, H.; Pandey, A.; Lane, M.D. Upstream stimulatory factors regulate the C/EBP α gene during differentiation of 3T3-L1 preadipocytes. *Biochem. Biophys. Res. Commun.,* **2007,** *354*(2), 517-521.
[http://dx.doi.org/10.1016/j.bbrc.2007.01.008] [PMID: 17239350]

[134] Patterson, A.D.; Li, H.; Eichler, G.S.; Krausz, K.W.; Weinstein, J.N.; Fornace, A.J., Jr; Gonzalez, F.J.; Idle, J.R. UPLC-ESI-TOFMS-based metabolomics and gene expression dynamics inspector self-organizing metabolomic maps as tools for understanding the cellular response to ionizing radiation. *Anal. Chem.,* **2008,** *80*(3), 665-674.
[http://dx.doi.org/10.1021/ac701807v] [PMID: 18173289]

[135] Xu, Y.; Cheung, W.; Winder, C.L.; Dunn, W.B.; Goodacre, R. Metabolic profiling of meat: assessment of pork hygiene and contamination with Salmonella typhimurium. *Analyst (Lond.),* **2011,** *136*(3), 508-514.
[http://dx.doi.org/10.1039/C0AN00394H] [PMID: 21113559]

[136] Khakimov, B.; Amigo, J.M.; Bak, S.; Engelsen, S.B. Plant metabolomics: resolution and quantification of elusive peaks in liquid chromatography-mass spectrometry profiles of complex plant extracts using multi-way decomposition methods. *J. Chromatogr. A,* **2012,** *1266*, 84-94.
[http://dx.doi.org/10.1016/j.chroma.2012.10.023] [PMID: 23107118]

[137]   Vaclavik, L.; Schreiber, A.; Lacina, O.; Cajka, T.; Hajslova, J. Liquid chromatography–mass spectrometry-based metabolomics for authenticity assessment of fruit juices. *Metabolomics,* **2012**, *8,* 793-803.
[http://dx.doi.org/10.1007/s11306-011-0371-7]

[138]   Ouyang, M.; Zhang, Z.; Chen, C.; Liu, X.; Liang, Y. Application of sparse linear discriminant analysis for metabolomics data. *Anal. Methods,* **2014**, *6,* 9037-9044.
[http://dx.doi.org/10.1039/C4AY01715C]

[139]   Phua, L.C.; Wilder-Smith, C.H.; Tan, Y.M.; Gopalakrishnan, T.; Wong, R.K.; Li, X.; Kan, M.E.; Lu, J.; Keshavarzian, A.; Chan, E.C.Y. Gastrointestinal Symptoms and Altered Intestinal Permeability Induced by Combat Training Are Associated with Distinct Metabotypic Changes. *J. Proteome Res.,* **2015**, *14*(11), 4734-4742.
[http://dx.doi.org/10.1021/acs.jproteome.5b00603] [PMID: 26506213]

[140]   Zhang, A.H.; Sun, H.; Han, Y.; Yan, G.L.; Yuan, Y.; Song, G.C.; Yuan, X.X.; Xie, N.; Wang, X.J. Ultraperformance liquid chromatography-mass spectrometry based comprehensive metabolomics combined with pattern recognition and network analysis methods for characterization of metabolites and metabolic pathways from biological data sets. *Anal. Chem.,* **2013**, *85*(15), 7606-7612.
[http://dx.doi.org/10.1021/ac401793d] [PMID: 23845028]

[141]   Li, Y.; Ju, L.; Hou, Z.; Deng, H.; Zhang, Z.; Wang, L.; Yang, Z.; Yin, J.; Zhang, Y. Screening, verification, and optimization of biomarkers for early prediction of cardiotoxicity based on metabolomics. *J. Proteome Res.,* **2015**, *14*(6), 2437-2445.
[http://dx.doi.org/10.1021/pr501116c] [PMID: 25919346]

[142]   Gao, R.; Cheng, J.; Fan, C.; Shi, X.; Cao, Y.; Sun, B.; Ding, H.; Hu, C.; Dong, F.; Yan, X. Serum metabolomics to identify the liver disease-specific biomarkers for the progression of hepatitis to hepatocellular carcinoma. *Sci. Rep.,* **2015**, *5,* 18175.
[http://dx.doi.org/10.1038/srep18175] [PMID: 26658617]

[143]   Jackson, J.E. *A user's guide to principal components*; John Wiley & Sons: New York, **2005**.

[144]   Jing, L.; Lei, Z.; Zhang, G.; Pilon, A.C.; Huhman, D.V.; Xie, R.; Xi, W.; Zhou, Z.; Sumner, L.W. Metabolite profiles of essential oils in citrus peels and their taxonomic implications. *Metabolomics,* **2015**, *11,* 952-963.
[http://dx.doi.org/10.1007/s11306-014-0751-x]

[145]   Kohonen, T. *Self-organizing maps, Springer series in information sciences, 15 Springer Science & Business Media 30*; , **1995**.

[146]   Bro, R.; Andersson, C.A.; Kiers, H.A. PARAFAC2—Part II. Modeling chromatographic data with retention time shifts. *J. Chemometr.,* **1999**, *13,* 295-309.
[http://dx.doi.org/10.1002/(SICI)1099-128X(199905/08)13:3/4<295::AID-CEM547>3.0.CO;2-Y]

[147]   Kiers, H.A.; Ten Berge, J.M.; Bro, R. PARAFAC2—Part I. A direct fitting algorithm for the PARAFAC2 model. *J. Chemometr.,* **1999**, *13,* 275-294.
[http://dx.doi.org/10.1002/(SICI)1099-128X(199905/08)13:3/4<275::AID-CEM543>3.0.CO;2-B]

[148]   Bro, R. PARAFAC. Tutorial and applications. *Chemom. Intell. Lab. Syst.,* **1997**, *38,* 149-171.
[http://dx.doi.org/10.1016/S0169-7439(97)00032-4]

[149]   Gorrochategui, E.; Casas, J.; Porte, C.; Lacorte, S.; Tauler, R. Chemometric strategy for untargeted lipidomics: biomarker detection and identification in stressed human placental cells. *Anal. Chim. Acta,* **2015**, *854,* 20-33.
[http://dx.doi.org/10.1016/j.aca.2014.11.010] [PMID: 25479864]

[150]   Barker, M.; Rayens, W. Partial least squares for discrimination. *J. Chemometr.,* **2003**, *17,* 166-173.
[http://dx.doi.org/10.1002/cem.785]

[151]   Trygg, J.; Wold, S. Orthogonal projections to latent structures (O☐PLS). *J. Chemometr.,* **2002**, *16,* 119-128.

[http://dx.doi.org/10.1002/cem.695]

[152]   Madsen, R.; Lundstedt, T.; Trygg, J. Chemometrics in metabolomics--a review in human disease diagnosis. *Anal. Chim. Acta,* **2010**, *659*(1-2), 23-33.
[http://dx.doi.org/10.1016/j.aca.2009.11.042] [PMID: 20103103]

[153]   Diémé, B.; Mavel, S.; Blasco, H.; Tripi, G.; Bonnet-Brilhault, F.; Malvy, J.; Bocca, C.; Andres, C.R.; Nadal-Desbarats, L.; Emond, P. Metabolomics Study of Urine in Autism Spectrum Disorders Using a Multiplatform Analytical Methodology. *J. Proteome Res.,* **2015**, *14*(12), 5273-5282.
[http://dx.doi.org/10.1021/acs.jproteome.5b00699] [PMID: 26538324]

[154]   Shawe-Taylor, J.; Cristianini, N. *Kernel methods for pattern analysis*; Cambridge University Press: Cambrige, England, **2004**.
[http://dx.doi.org/10.1017/CBO9780511809682]

[155]   Vapnik, V. *The nature of statistical learning theory*; Springer Science & Business Media: Berlin, **2013**.

[156]   Walczak, B.; Massart, D.L. The Radial Basis Functions — Partial Least Squares approach as a flexible non-linear regression technique. *Anal. Chim. Acta,* **1996**, *331*, 177-185.
[http://dx.doi.org/10.1016/0003-2670(96)00202-4]

[157]   Luts, J.; Ojeda, F.; Van de Plas, R.; De Moor, B.; Van Huffel, S.; Suykens, J.A.K. A tutorial on support vector machine-based methods for classification problems in chemometrics. *Anal. Chim. Acta,* **2010**, *665*(2), 129-145.
[http://dx.doi.org/10.1016/j.aca.2010.03.030] [PMID: 20417323]

[158]   Huang, J-H.; Fu, L.; Li, B.; Xie, H-L.; Zhang, X.; Chen, Y.; Qin, Y.; Wang, Y.; Zhang, S.; Huang, H.; Liao, D.; Wang, W. Distinguishing the serum metabolite profiles differences in breast cancer by gas chromatography mass spectrometry and random forest method. *RSC Advances,* **2015**, *5*, 58952-58958.
[http://dx.doi.org/10.1039/C5RA10130A]

[159]   Stone, M. *Cross-validatory choice and assessment of statistical predictions, J. Royal Stat. Soc, Series B (Method.)*; DOI, **1974**, pp. 111-147.

[160]   Krstajic, D.; Buturovic, L.J.; Leahy, D.E.; Thomas, S. Cross-validation pitfalls when selecting and assessing regression and classification models. *J. Cheminform.,* **2014**, *6*(1), 10.
[http://dx.doi.org/10.1186/1758-2946-6-10] [PMID: 24678909]

[161]   Geisser, S. **1975**, 320-328.

[162]   Shao, J. Linear Model Selection by Cross-validation. *J. Am. Stat. Assoc.,* **1993**, *88*, 486-494.
[http://dx.doi.org/10.1080/01621459.1993.10476299]

[163]   Westerhuis, J.A.; Hoefsloot, H.C.J.; Smit, S. D.J. Vis, A.K. Smilde, E.J.J. van Velzen, J.P.M. van Duijnhoven, F.A. van Dorsten, Assessment of PLSDA cross validation. *Metabolomics,* **2008**, *4*, 81-89.
[http://dx.doi.org/10.1007/s11306-007-0099-6]

[164]   Eidhammer, I.; Flikka, K.; Martens, L.; Mikalsen, S-O. *Computational methods for mass spectrometry proteomics*; John Wiley & Sons: New York, USA, **2008**.

[165]   Colinge, J.; Bennett, K.L. Introduction to computational proteomics. *PLOS Comput. Biol.,* **2007**, *3*(7): e114.
[http://dx.doi.org/10.1371/journal.pcbi.0030114] [PMID: 17676979]

[166]   Matthiesen, R. *Mass spectrometry data analysis in proteomics*; Springer Science & Business Media: Berlin, **2007**.

[167]   Kumar, C.; Mann, M. Bioinformatics analysis of mass spectrometry-based proteomics data sets. *FEBS Lett.,* **2009**, *583*(11), 1703-1712.
[http://dx.doi.org/10.1016/j.febslet.2009.03.035] [PMID: 19306877]

[168]   Gilany, K.; Moens, L.; Dewilde, S. Mass spectrometry based proteomics in the life sciences: a review. *J. Paramed. Sci.,* **2010**, *1*, 53-78.

[169] Urfer, W.; Grzegorczyk, M.; Jung, K. Statistics for proteomics: a review of tools for analyzing experimental data. *Proteomics,* **2006**, *6* Suppl. 2, 48-55.
[http://dx.doi.org/10.1002/pmic.200600554] [PMID: 17031797]

[170] Malmström, J.; Lee, H.; Aebersold, R. Advances in proteomic workflows for systems biology. *Curr. Opin. Biotechnol.,* **2007**, *18*(4), 378-384.
[http://dx.doi.org/10.1016/j.copbio.2007.07.005] [PMID: 17698335]

[171] Cox, J.; Mann, M. MaxQuant enables high peptide identification rates, individualized p.p.b.-range mass accuracies and proteome-wide protein quantification. *Nat. Biotechnol.,* **2008**, *26*(12), 1367-1372.
[http://dx.doi.org/10.1038/nbt.1511] [PMID: 19029910]

[172] Teleman, J.; Dowsey, A.W.; Gonzalez-Galarza, F.F.; Perkins, S.; Pratt, B.; Röst, H.L.; Malmström, L.; Malmström, J.; Jones, A.R.; Deutsch, E.W.; Levander, F. Numerical compression schemes for proteomics mass spectrometry data. *Mol. Cell. Proteomics,* **2014**, *13*(6), 1537-1542.
[http://dx.doi.org/10.1074/mcp.O114.037879] [PMID: 24677029]

[173] Patterson, S.D.; Aebersold, R.H. Proteomics: the first decade and beyond. *Nat. Genet.,* **2003**, *33* Suppl., 311-323.
[http://dx.doi.org/10.1038/ng1106] [PMID: 12610541]

[174] Aebersold, R.; Mann, M. Mass spectrometry-based proteomics. *Nature,* **2003**, *422*(6928), 198-207.
[http://dx.doi.org/10.1038/nature01511] [PMID: 12634793]

[175] Johnson, R.S.; Taylor, J.A. Searching sequence databases via de novo peptide sequencing by tandem mass spectrometry. *Mol. Biotechnol.,* **2002**, *22*(3), 301-315.
[http://dx.doi.org/10.1385/MB:22:3:301] [PMID: 12448884]

[176] Mann, M.; Wilm, M. Error-tolerant identification of peptides in sequence databases by peptide sequence tags. *Anal. Chem.,* **1994**, *66*(24), 4390-4399.
[http://dx.doi.org/10.1021/ac00096a002] [PMID: 7847635]

[177] Eng, J.K.; McCormack, A.L.; Yates, J.R. An approach to correlate tandem mass spectral data of peptides with amino acid sequences in a protein database. *J. Am. Soc. Mass Spectrom.,* **1994**, *5*(11), 976-989.
[http://dx.doi.org/10.1016/1044-0305(94)80016-2] [PMID: 24226387]

[178] Clauser, K.R.; Baker, P.; Burlingame, A.L. Role of accurate mass measurement (+/- 10 ppm) in protein identification strategies employing MS or MS/MS and database searching. *Anal. Chem.,* **1999**, *71*(14), 2871-2882.
[http://dx.doi.org/10.1021/ac9810516] [PMID: 10424174]

[179] Zhang, X.; Asara, J.M.; Adamec, J.; Ouzzani, M.; Elmagarmid, A.K. Data pre-processing in liquid chromatography-mass spectrometry-based proteomics. *Bioinformatics,* **2005**, *21*(21), 4054-4059.
[http://dx.doi.org/10.1093/bioinformatics/bti660] [PMID: 16150809]

[180] Meng, Z.; Simmons-Willis, T.A.; Limbach, P.A. The use of mass spectrometry in genomics. *Biomol. Eng.,* **2004**, *21*(1), 1-13.
[http://dx.doi.org/10.1016/j.bioeng.2003.08.001] [PMID: 14715314]

[181] Dubitzky, W.; Granzow, M.; Berrar, D.P. *Fundamentals of data mining in genomics and proteomics*; Springer Science & Business Media: Berlin, **2007**.
[http://dx.doi.org/10.1007/978-0-387-47509-7]

[182] Grahn, H.; Geladi, P. *Techniques and applications of hyperspectral image analysis*; John Wiley & Sons: New York, USA, **2007**.
[http://dx.doi.org/10.1002/9780470010884]

[183] Salzer, R.; Siesler, H.W. *Infrared and Raman spectroscopic imaging*; John Wiley & Sons: New York, USA, **2007**.

[184] Liu, D.; Zeng, X-A.; Sun, D-W. NIR spectroscopy and imaging techniques for evaluation of fish

quality—a review. *Appl. Spectrosc. Rev.,* **2013**, *48*, 609-628.
[http://dx.doi.org/10.1080/05704928.2013.775579]

[185] Gowen, A.A.; O'Donnell, C.P.; Cullen, P.J.; Bell, S.E. Recent applications of Chemical Imaging to pharmaceutical process monitoring and quality control. *Eur. J. Pharm. Biopharm.,* **2008**, *69*(1), 10-22.
[http://dx.doi.org/10.1016/j.ejpb.2007.10.013] [PMID: 18164926]

[186] Kherlopian, A.R.; Song, T.; Duan, Q.; Neimark, M.A.; Po, M.J.; Gohagan, J.K.; Laine, A.F. A review of imaging techniques for systems biology. *BMC Syst. Biol.,* **2008**, *2*, 74.
[http://dx.doi.org/10.1186/1752-0509-2-74] [PMID: 18700030]

[187] Van der Meer, F.D.; Van der Werff, H.M.; Van Ruitenbeek, F.J.; Hecker, C.A.; Bakker, W.H.; Noomen, M.F.; Van Der Meijde, M.; Carranza, E.J.M.; De Smeth, J.B.; Woldai, T. Multi-and hyperspectral geologic remote sensing: A review. *Int. J. Appl. Earth Obs. Geoinf.,* **2012**, *14*, 112-128.
[http://dx.doi.org/10.1016/j.jag.2011.08.002]

[188] Setou, M. *Imaging mass spectrometry: protocols for mass microscopy*; Springer Science & Business Media: Berlin, Germany, **2010**.
[http://dx.doi.org/10.1007/978-4-431-09425-8]

[189] Rubakhin, S.S.; Sweedler, J.V. *Mass spectrometry imaging: Principles and protocols*; Springer Science & Business Media: Berlin, Germany, **2010**.
[http://dx.doi.org/10.1007/978-1-60761-746-4]

[190] Louie, K.B.; Bowen, B.P.; McAlhany, S.; Huang, Y.; Price, J.C.; Mao, J.H.; Hellerstein, M.; Northen, T.R. Mass spectrometry imaging for in situ kinetic histochemistry. *Sci. Rep.,* **2013**, *3*, 1656.
[http://dx.doi.org/10.1038/srep01656] [PMID: 23584513]

[191] Chughtai, K.; Heeren, R.M. Mass spectrometric imaging for biomedical tissue analysis. *Chem. Rev.,* **2010**, *110*(5), 3237-3277.
[http://dx.doi.org/10.1021/cr100012c] [PMID: 20423155]

[192] McDonnell, L.A.; Heeren, R.M. Imaging mass spectrometry. *Mass Spectrom. Rev.,* **2007**, *26*(4), 606-643.
[http://dx.doi.org/10.1002/mas.20124] [PMID: 17471576]

[193] Amigo, J.M. Practical issues of hyperspectral imaging analysis of solid dosage forms. *Anal. Bioanal. Chem.,* **2010**, *398*(1), 93-109.
[http://dx.doi.org/10.1007/s00216-010-3828-z] [PMID: 20496027]

[194] Esbensen, K.; Geladi, P. Strategy of multivariate image analysis (MIA). *Chemom. Intell. Lab. Syst.,* **1989**, *7*, 67-86.
[http://dx.doi.org/10.1016/0169-7439(89)80112-1]

[195] Geladi, P.; Isaksson, H.; Lindqvist, L.; Wold, S.; Esbensen, K. Principal component analysis of multivariate images. *Chemom. Intell. Lab. Syst.,* **1989**, *5*, 209-220.
[http://dx.doi.org/10.1016/0169-7439(89)80049-8]

[196] de Juan, A.; Tauler, R.; Dyson, R.; Marcolli, C.; Rault, M.; Maeder, M. Spectroscopic imaging and chemometrics: a powerful combination for global and local sample analysis. *Trends Analyt. Chem.,* **2004**, *23*, 70-79.
[http://dx.doi.org/10.1016/S0165-9936(04)00101-3]

[197] Marro, M.; Taubes, A.; Abernathy, A.; Balint, S.; Moreno, B.; Sanchez-Dalmau, B.; Martínez-Lapiscina, E.H.; Amat-Roldan, I.; Petrov, D.; Villoslada, P. Dynamic molecular monitoring of retina inflammation by *in vivo* Raman spectroscopy coupled with multivariate analysis. *J. Biophotonics,* **2014**, *7*(9), 724-734.
[http://dx.doi.org/10.1002/jbio.201300101] [PMID: 24019106]

[198] Piqueras, S.; Duponchel, L.; Tauler, R.; de Juan, A. Monitoring polymorphic transformations by using in situ Raman hyperspectral imaging and image multiset analysis. *Anal. Chim. Acta,* **2014**, *819*, 15-25.
[http://dx.doi.org/10.1016/j.aca.2014.02.027] [PMID: 24636406]

[199] Zhang, X.; Tauler, R. Application of multivariate curve resolution alternating least squares (MCR-ALS) to remote sensing hyperspectral imaging. *Anal. Chim. Acta*, **2013**, *762*, 25-38.
[http://dx.doi.org/10.1016/j.aca.2012.11.043] [PMID: 23327942]

[200] Jaumot, J.; Tauler, R. Potential use of multivariate curve resolution for the analysis of mass spectrometry images. *Analyst (Lond.)*, **2015**, *140*(3), 837-846.
[http://dx.doi.org/10.1039/C4AN00801D] [PMID: 25460200]

[201] Bedia, C.; Tauler, R.; Jaumot, J. Analysis of multiple mass spectrometry images from different Phaseolus vulgaris samples by multivariate curve resolution. *Talanta*, **2017**, *175*, 557-565.
[http://dx.doi.org/10.1016/j.talanta.2017.07.087] [PMID: 28842033]

[202] Spengler, B. Mass spectrometry imaging of biomolecular information. *Anal. Chem.*, **2015**, *87*(1), 64-82.
[http://dx.doi.org/10.1021/ac504543v] [PMID: 25490190]

[203] Alexandrov, T. MALDI imaging mass spectrometry: statistical data analysis and current computational challenges. *BMC Bioinformatics*, **2012**, *13* Suppl. 16, S11.
[http://dx.doi.org/10.1186/1471-2105-13-S16-S11] [PMID: 23176142]

[204] Bartels, A.; Dülk, P.; Trede, D.; Alexandrov, T.; Maaß, P. Compressed sensing in imaging mass spectrometry. *Inverse Probl.*, **2013**, *29*: 125015.
[http://dx.doi.org/10.1088/0266-5611/29/12/125015]

[205] Cumpson, P.J.; Fletcher, I.W.; Sano, N.; Barlow, A.J. Rapid multivariate analysis of 3D ToF⬜SIMS data: graphical processor units (GPUs) and low⬜discrepancy subsampling for large⬜scale principal component analysis. *Surf. Interface Anal.*, **2016**, *48*, 1328-1336.
[http://dx.doi.org/10.1002/sia.6042]

[206] Galli, M.; Zoppis, I.; Smith, A.; Magni, F.; Mauri, G. Machine learning approaches in MALDI-MSI: clinical applications. *Expert Rev. Proteomics*, **2016**, *13*(7), 685-696.
[http://dx.doi.org/10.1080/14789450.2016.1200470] [PMID: 27322705]

[207] Källback, P.; Nilsson, A.; Shariatgorji, M.; Andrén, P.E. MsIQuant–quantitation software for mass spectrometry imaging enabling fast access, visualization, and analysis of large data sets. *Anal. Chem.*, **2016**, *88*(8), 4346-4353.
[http://dx.doi.org/10.1021/acs.analchem.5b04603] [PMID: 27014927]

[208] Race, A.M.; Palmer, A.D.; Dexter, A.; Steven, R.T.; Styles, I.B.; Bunch, J. SpectralAnalysis: Software for the Masses. *Anal. Chem.*, **2016**, *88*(19), 9451-9458.
[http://dx.doi.org/10.1021/acs.analchem.6b01643] [PMID: 27558772]

[209] Tautenhahn, R.; Böttcher, C.; Neumann, S. Highly sensitive feature detection for high resolution LC/MS. *BMC Bioinformatics*, **2008**, *9*, 504.
[http://dx.doi.org/10.1186/1471-2105-9-504] [PMID: 19040729]

[210] Stolt, R.; Torgrip, R.J.; Lindberg, J.; Csenki, L.; Kolmert, J.; Schuppe-Koistinen, I.; Jacobsson, S.P. Second-order peak detection for multicomponent high-resolution LC/MS data. *Anal. Chem.*, **2006**, *78*(4), 975-983.
[http://dx.doi.org/10.1021/ac050980b] [PMID: 16478086]

[211] Bedia, C.; Tauler, R.; Jaumot, J. Compression strategies for the chemometric analysis of mass spectrometry imaging data. *J. Chemometr.*, **2016**, *30*, 575-588.
[http://dx.doi.org/10.1002/cem.2821]

[212] Siano, G.G.; Pérez, I.S.; García, M.D.G.; Galera, M.M.; Goicoechea, H.C. Multivariate curve resolution modeling of liquid chromatography-mass spectrometry data in a comparative study of the different endogenous metabolites behavior in two tomato cultivars treated with carbofuran pesticide. *Talanta*, **2011**, *85*(1), 264-275.
[http://dx.doi.org/10.1016/j.talanta.2011.03.064] [PMID: 21645698]

CHAPTER 4

Recent Developments of Allied Techniques of Qualitative Analysis of Heavy Metal Ions in Aqueous Solutions with Special Reference to Modern Mass Spectrometry

Bijoy Sankar Boruah and **Rajib Biswas**[*]

Applied Optics and Photonics Research Laboratory, Tezpur University, Napaam-784028, Assam, India

Abstract: Heavy metal ions are basic elements of earth crust. These metal ions are non-biodegradable in nature and tend to accumulate in our ecosystem in due course of time. Some of the most toxic heavy metal ions include arsenic, mercury, cadmium, lead, nickel *etc.* The toxicity level depends on density for any biological system. Due to increasing applications of heavy metal ion compounds in industrial, agricultural and medical fields, water pollution induced by excess levels of heavy metal ion becomes a big crisis for us. As such, detection of heavy metal ions in water is an important issue for us. Mass spectroscopy methods are the most conventionally applied methods for the detection of heavy metal ions in water. Some of the mass spectroscopic methods are atomic absorption spectroscopy, inductively coupled plasma mass spectroscopy, graphite furnace atomic absorption spectroscopy *etc.* These methods have well detection capability of heavy metal ions in water with good selectivity and sensitivity. Along with mass spectroscopic methods, the use of optical fiber technology for heavy metal ions detection is remarkable. Optical fiber based sensors system for the detection of heavy metal ions basically works by changing the effective refractive index of its surroundings. For selective binding of heavy metal ions, sensitive layers are coated on optical fiber probe. Laser or light emitting diode is used as a light source in an optical fiber sensor for signal purpose. Accordingly, output response for various heavy metal ions is recorded on an optical spectrometer. From their output response, we can determine the concentration of metal ions present in water. It is noticed that optical fiber sensor can also have good sensitivity and selectivity towards the detection of heavy metal ions as mass spectroscopy methods.

Keywords: Arsenic, Cadmium, Colorimetric, Detection, Electrochemical, Heavy metal ion, Lead, Mass spectrometry, Mercury, Optical fiber sensor, Sources of heavy metal ions.

[*] **Corresponding author Rajib Biswas:** Applied Optics and Photonics Research Laboratory, Tezpur University, Napaam-784028, Assam, India; Tel/Fax: +91-3712-267005/6; E-mail: rajib@tezu.ernet.in

INTRODUCTION

Heavy metals are the natural elements of the earth crust. The term "heavy" refers to relatively high density and atomic mass. This means that heavy metals possess high mass to volume ratio. The metals having atomic weights in between 63.5 and 200.6 g/mol with density 5 g /cm^3 can be labeled as heavy metals [1, 2]. But it is worthy to mention that any metal can be called heavy metal if its' presence in any system crosses the limit of its permissible limit irrespective of atomic mass or density. When the presence of any metal in any system exceeds its permissible limit, then it becomes toxic. Hence, it can be implicitly stated that heavy metals are toxic for any biological system for longer endurance [3, 4].

Metalloid such as arsenic is also called a heavy metal because it emerges to be highly toxic at a very low level of concentration [5, 6]. There are other metals such as iron, copper, manganese, zinc, cobalt *etc*. They are regarded as micro-nutrients of biological system. These micro-nutrients facilitate lots of activities in human body and also in plant. In case, the concentration of these metals increases beyond the permissible limit; they become poisonous and may cause various health issues [7, 8]. The most commonly known heavy metal ions are mercury (Hg), lead (Pb), cadmium (Cd), arsenic (As), chromium (Cr) and nickel (Ni). Even at very low concentration such as parts per billion (ppb) *i.e.* µg/L, these metal ions are harmful and intake or inhalation may cause different types of ailments [9].

In general, heavy metals are ubiquitously found. Nowadays, anthropogenic sources have become a major concern for us. Heavy metal compounds are used in different industrial sectors, and in agricultural field as pesticides, fertilizers, *etc*. Similarly, in semiconductor industry, arsenic is used as a doping element. After utilization of these heavy metal compounds, they finally enter environment [10, 11]. Being non-biodegradable in nature, these metal ions accumulate in our ecosystem and contaminate aquatic bodies eventually. As it is known that water is a primary essential element of the food cycle, hence utilization of heavy metal polluted water leads to contamination of food cycle. Through drinking of contaminated water or food, these heavy metals enter our body. These heavy metal ions have a tendency to bind with the thiol group of proteins [12]. In this way, these heavy metal ions are bio accumulated in different organisms of human body. Due to this bio accumulation, various types of diseases such as cancer, nervous system damage, kidney, liver, lung problem, high blood pressure *etc*. may occur [13].

There are several organizations such as the World Health Organization (WHO), the Environment Protection Agency (EPA), and the Centre for Disease Control

(CDC) for monitoring the effect of these metal ions to our environment. In Fig. (**1a**) and Fig. (**1b**), we have shown the most commonly found heavy metal ions and their various sources, respectively. These heavy metal ions are called "environment health hazards" depending on their toxicity and contamination in air, water and soil [9, 14]. In Table **1**, the most toxic heavy metal ions along with WHO permissible limits and their effect on human body are enlisted.

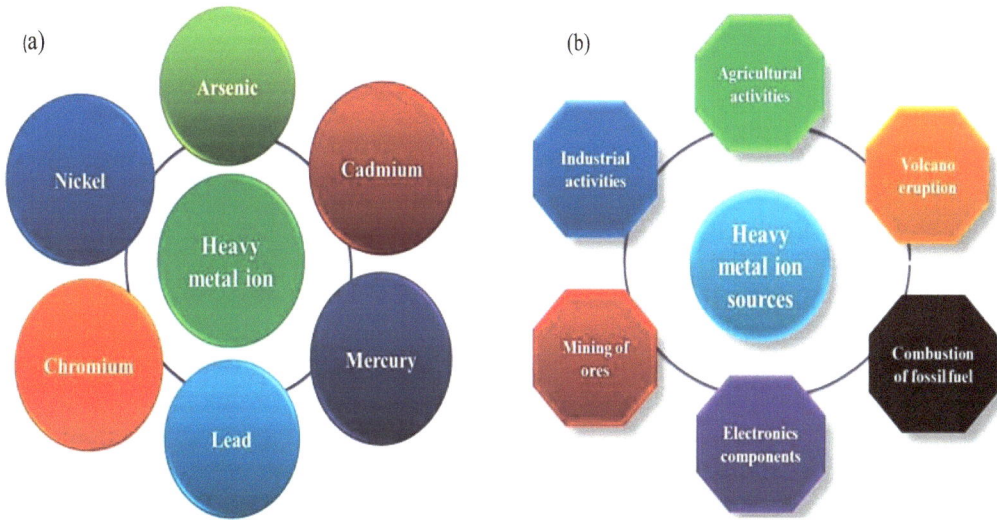

Fig. (**1**). (a) The most toxic heavy metal ions and (b) various sources of heavy metal ions.

Table 1. Toxic heavy metal ions, their sources, WHO permissible limit and their various health effects.

Heavy Metal	Sources	WHO Limits (ppb)	Health Effects	References
Mercury (Hg)	Volcanic eruptions, gold mining, burning of fossil fuels	2	Damage of kidney, heart, muscle, respiratory system, liver.	[15]
Arsenic (As)	Electronics industry, agricultural use of pesticides, fertilizer	10	Skin cancer, neurological damage, high blood pressure, kidney, reproductive system damage.	[16]
Cadmium (Cd)	Plastic materials, paints, pigments, photovoltaic cells	5	Cell damage, weight loss, metabolic disorder, renal cancer.	[17]
Lead (Pb)	Plastic pipe, batteries, paints pigments.	10	Anemia, kidney damage, memory loss, hepatopathy, neurological problem.	[18]

(Table 1) cont.....

Heavy Metal	Sources	WHO Limits (ppb)	Health Effects	References
Nickel (Ni)	Industrial and clinical application of nickel compounds	20	Lung cancer, DNA damage, kidney damage, allergy.	[19]
Chromium (Cr)	Chrome plating steel manufacture, leather tanning, and pigment production	50	Genetic disorder, kidney and lung cancer, skin damage.	[20, 21]

It is evident from this table that heavy metal ions may arise from various sources of manmade activities. For different heavy metal ions, the World health Organization sets different permissible limits depending on their toxicity. It is found that some heavy metal ions are very toxic at a very low concentration. For example, the permissible limit for mercury is 2 ppb, above which it is considered as toxic for the environment. Likewise, allowed level of cadmium is 5 ppb. Similarly, the permissible limit, for arsenic and lead is set to be 10 ppb. Therefore, monitoring of these heavy metal ions in environment is very important. Taking the enormity of ill-effects of heavy metal ion contaminated aqueous solutions into account, it is imperative that their concentration should be monitored through effective mechanism. In this direction, there are several techniques to achieve this objective. The purpose of this chapter is to appraise readers about these spectroscopic methods such as mass spectroscopy and its allied techniques accompanied by their comparative analysis.

TECHNIQUES FOR SENSING HEAVY METAL ION

For the analysis of heavy metal ions in environment, there are several well-established techniques such as colorimetric, electrochemical, mass spectrometry and optical fiber technique. These methods are endowed with good sensitivity as well as selectivity towards metal ion detection in our environment. Accordingly, the recent progress of these techniques in detection of heavy metal ion detection in aqueous medium is illustrated in the following sections.

Colorimetric Technique

For detection of heavy metal ions, colorimetric method is a simple and low cost method which can be visualized through the naked eye. Colorimetric technique is a light interaction based technique with colored compound. It is a method to measure the density of a colored compound in solution. The advantage of colorimetric method is that it can be assessed by the naked eye. Colorimetric method is used in various fields like as medical application, food chemistry, agricultural field and environment monitoring [22, 23]. In environmental science, monitoring of water quality is an important issue. Colorimetric method can be applied for monitoring of water quality. With the help of colorimetric method, we can easily know the presence of heavy metal ions in water.

Since colorimetric approach is based on color of analytes, hence a reagent is required to produce color in solution. In colorimetric detection of heavy metal ions, utilization of nanomaterial is increasing gradually. This is because nanomaterials have different properties from their bulk constituents. Depending on the size and shape, nanoparticles exhibit different color in visible region. The most widely used nanoparticles in heavy metal ion detection are silver and gold. This wide use can be ascribed to the presence of high optical extinction coefficient along with high sensitivity [14, 24].

Here in Fig. (**2a**), we have shown the colorimetric detection of arsenic with the help of functionalized gold nanoparticles as sensing unit. Fig. (**2b**) shows the interaction mechanism of arsenic with functionalized gold nanoparticles for which we get color change from win red to blue. As the gold nanoparticles have changed their color from wine red to blue color, this implies that optical response of these nanoparticles has altered. This change in optical response is shown in Fig. (**2c**) and (**2d**). For the synthesis of silver or gold nanoparticles, chemical reduction method or synthesis through green route can be applied. In chemical reduction method, tri sodium citrate, sodium boro hydrate, *etc.* are used as reducing agents. On other hand in green route, some nontoxic materials such as chitosan, some leaf extracts *etc.* are used as reducing agent. Sometimes, some functionalizing agent such as glutathione, lysine *etc.* are also used to increase selectivity or sensitivity. As the heavy metal ion interacts with these nanomaterials; aggregation of nanoparticles occurs due to interaction. Owing to change in size and shape of the nanomaterial in aggregation, these nanomaterials start displaying different colors. Depending on change in color, we can detect heavy metal ions in aqueous medium [25]. Detection of heavy metal ions using nanomaterial is basically a spectroscopic method. Here, UV-vis. spectrometer is generally used to measure the absorption spectra of nanoparticles. In this detection method, a calibration curve is taken using blank sample. After that, optical response of nanoparticle is measured. Then different concentrations of heavy metal ions are added to the nanoparticles and their optical response is measured. Finally, the difference between the optical response of heavy metal without nanoparticles and heavy metal ion containing nanoparticles will lead to the concentration of heavy metal ion present in that solution. In the Table **2** provided below, we have shown some recently published work based on colorimetric detection for heavy metal ion.

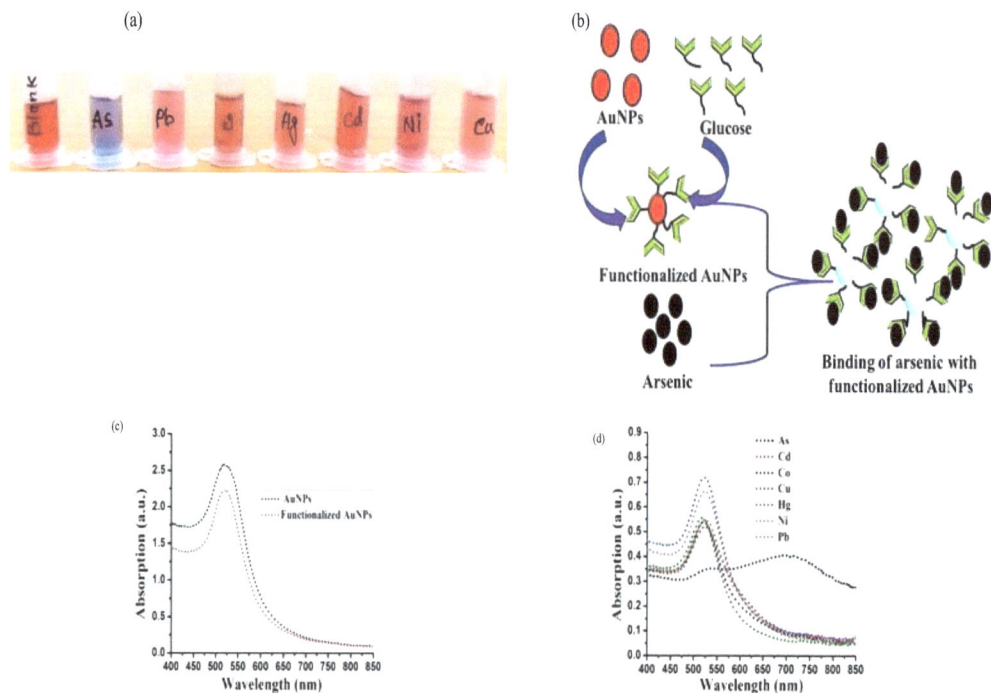

Fig. (2). Colorimetric detection of (a) arsenic with different heavy metal ions (b) interaction mechanism of arsenic with functionalized gold nanoparticles (c) Optical response of bare gold and functionalized gold nanoparticles and (d) Optical response of different heavy metal ions in functionalized gold nanoparticles in visible region [34].

Table 2. Colorimetric detection of various heavy metal ions using functionalizing nanoparticles, linear range with limit of detection.

Sensing Unit	Functionalizing Agent	Heavy Metal Ion	Linear Range	Limit of Detection	References
AuNPs	Glutathione	Lead	10^{-4}-10 μM	5×10^{-5}μM	[26]
AuNPs	Malic acid	Lead	0-10ppb	0.5ppb	[27]
AuNPs	Gallic acid	Lead	10.0–1000.0 nM	10 nM	[28]
AgNPs	Iminodiacetic acid	Lead	0.4 t- 8.0 μM	13 nM	[29]
AgNPs	1-(2-mercaptoethyl)-1,3,5-triazinane-2,4,6-trione	Lead	0.1-0.6μg/ml	0.02 and 0.06 μg/ml	[30]
AgNPs	Polyethylene glycol	Arsenic	5-13ppb	1ppb	[31]
AgNPs	Aptamer	Arsenic	50 - 700 ppb	6ppb	[32]

(Table 2) cont.....

Sensing Unit	Functionalizing Agent	Heavy Metal Ion	Linear Range	Limit of Detection	References
AuNPs	Polyethylene glycol	Arsenic	5–20 ppb	5ppb	[33]
AuNPs	Glucose	Arsenic	1–14 ppb	0.53 ppb	[34]
AuNPs	GSH-DTT-CYs-PDCA	Arsenic	2–20 ppb	2.5ppb	[35]
AuNPs	*Mangifera indica* leaf extract	Arsenic	1–11 ppb	1.2 ppb	[36]
AuNPs	Lysine	Mercury	1-1000 nM	2.9 nM	[37]
AuNPs	Thymine	Mercury	25–750 nM	50 nM	[38]
AgNPs	Thiamine	Mercury	0.01– 5 µM	5 nM	[39]
AuNPs	DNA	Mercury	-	20 ppb	[40]
AuNPs	Calixarene	Mercury	0.2 -100 µM	40 ppb	[41]
AuNPs	Thioctic acid	Mercury	10 - 20 µM	10 nM	[42]
AuNPs	Chitosan dithiocarbamate	Cadmium	50–500 µM	63 nM	[43]
AuNPs	di-(1H-pyrrol-2-yl)methanethione	Cadmium	0.5 - 16 µM	16.6 nM	[44]
AuNPs	6mercaptonicotinic acid and L-Cysteine	Cadmium	0.2-1.7µM		[45]
AgNPs	5-sulfosalicylic acid	Cadmium	0.05–1.1 µM	3nM	[46]
AgNPs/AuNPs	L-cysteine	Cadmium	0.4- 38.6 µM	44nM	[47]
AgNPs	Glutathione	Nickel	5–300 nM	5nM	[48]
AuNPs	Zwitterionic peptide	Nickel	60 – 160 nM	34nM	[49]
AgNPs	adenosine monophosphate - sodium dodecyl sulfonate	Nickel	4- 60 µM	0.60 µM	[50]
AgNPs	Glutathione and L-cysteine	Nickel	10-150 ppb	7.02 ppb	[51]
AgNPs	N-acetyl-L-cysteine	Nickel	2 - 48 µM	0.23 µM.	[52]

Optical Fiber Technique

An optical fiber is a waveguide that guides light along its length by total internal reflection. Optical fibers are generally used for communication and data transfer purpose. The guiding of light is basically dependent on the refractive indices of core and cladding of fiber. For propagation of light through the fiber, there are two conditions which are as follows:

1. Refractive index of core (n_1) must be greater than the refractive index (n_2) of the cladding.
2. The angle of incidence (θ) must be greater than the critical angle (θc).

Light propagation through the fiber also depends on the light launching angle to the fiber axis. There is an angle known as acceptance angle above which no light passes through the fiber. So, lager acceptance angle makes it easier to launch light into the fiber. This light gathering capacity of an optical fiber depends on numerical aperture which again relies on the refractive index of the fiber. When light is propagated through an optical fiber, it is characterized by different types of modes [53]. Modes are the possible no. of allowed paths. The modes that propagate at angles closer to critical angle of propagation are known as higher order modes and the modes that propagate with an angle lager than the critical angle of propagation is known as lower order modes. In case of lower order modes, light passes near the center of the fiber. But in higher order modes, electric fields are distributed more towards the edge of the wave guide and tend to send light energy into the cladding. We can increase the number of modes by increasing the core refractive index of the fiber, *i.e.* the numerical aperture of the fiber. Lager the numerical aperture, we will get higher order modes which allow field distribution near the cladding of the fiber [54]. This is beneficial for us because we can use it for making sensor. If we remove a small portion of cladding, then the fields are in the outer part of the core of fiber and they interact with surrounding environment. Any kind of external agent has its own refractive index. So, changes in refractive index also change the output of the fiber. The electromagnetic field residing on declared portion of the fiber is known as evanescent wave. Depending on this evanescent wave, scientist and researchers build different types of sensor for measuring many things [55, 56].

In the exposed cladding portion, we use thin metal films or nanoparticles to increase the sensitivity of optical fiber sensor. Depending on it, there are two types of optical fiber sensors namely, surface plasmon resonance (SPR) and localized surface Plasmon resonance (LSPR) based sensors. In case of surface plasmon resonance based sensors, we use a thin metal film on the exposed portion. SPR surface Plasmon can be divided into two parts, one is symmetric surface plasmon and another is anti-symmetric surface plasmon [57]. In symmetric surface plasmon, propagation constant and the attenuation constant increase with increasing metal film thickness. But in anti-symmetric SPR, it decreases with increasing metal thickness. On the other hand, deposition or coating of nanoparticles on exposed portion results in in localized surface resonance (LSPR) based sensors. LSPR based sensors have many advantages as compared to the SPR sensors. LSPR sensors are highly sensitive and label free technology that can be controlled by size, shape, material of nanoparticles [58].

Nowadays, optical fiber is used to make sensor due to its small size, light weight, immunity to electromagnetic interference (EMI), high temperature performance, large bandwidth, high sensitivity, and environmental ruggedness. Optical fiber sensors are used for temperature, pressure, strain, humidity, pH, gas, biomolecules and chemicals sensing [59]. Depending on the shape of optical fiber, sensors may be straight, tapered, D-shaped, fiber-grating and U-bent sensors.

Optical fiber sensors have also found applications for heavy metal detection in water. In the recent years, many research groups have developed different optical fiber sensors for monitoring different heavy metal ions in aqueous medium. These optical fiber sensors are basically U-shaped, tapered, fiber-grating and straight one. Recently, Biswas and his research team have reported an optical fiber sensor unit for detection of lead ion in aqueous medium. They have used a U-shaped optical fiber probe as sensing unit and coated with sensitive materials as sensing medium [60, 61]. Preparation of U-shaped optical probe and surface modification of U-shaped probe are shown in Fig. (**3a and b**). Surface is modified with chitosan–glutathione composite. Optical sensing system and interaction of lead with sensor probe is shown in Fig. (**3c and 3d**). Among other heavy metal ions, optical response of lead ion is found high that implies the selectivity towards lead ions as shown in Fig. (**3e**). The output response for various lead ion solutions in ppb range is shown in Fig. (**3f**). As the lead ion interacts with this sensitive material, the output intensity of the sensor unit is modulated. This modulated light intensity is measured in terms of voltage, power and intensity using an optical photo detector with respect to time for various concentration of lead ion. They have found that the sensor unit has yielded limit of detection 1.3 ppb and 1.75 ppb which is well below the WHO permissible limit 10 ppb [61]. Similarly, glucose functionalized silver nanoparticles coated on U-shaped optical fiber probe have also been reported for detection of mercury by Mukherji and his research team. The limit of detection of this sensor system is 2 ppb [62]. Here in the Table 3, we have enlisted the recent development of optical fiber sensors for detection of heavy metal ions in terms of sensing probe, sensitive materials linear range and limit of detection.

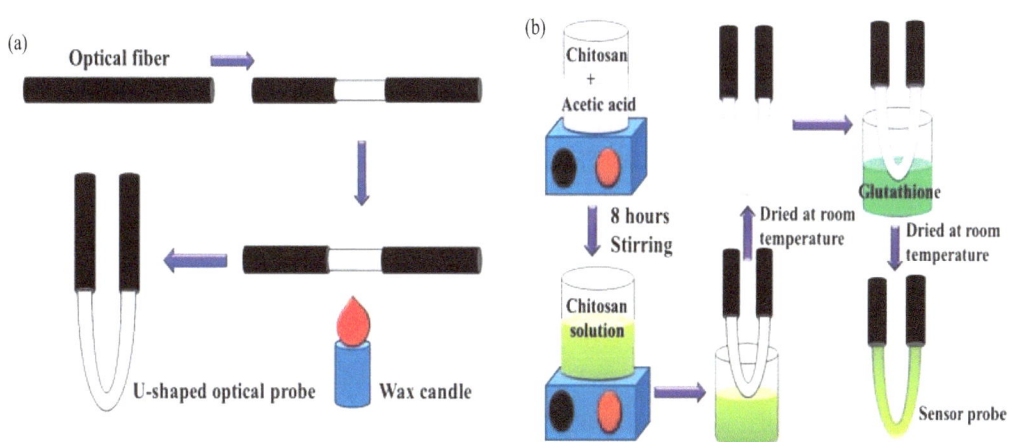

Fig. 3 cont.....

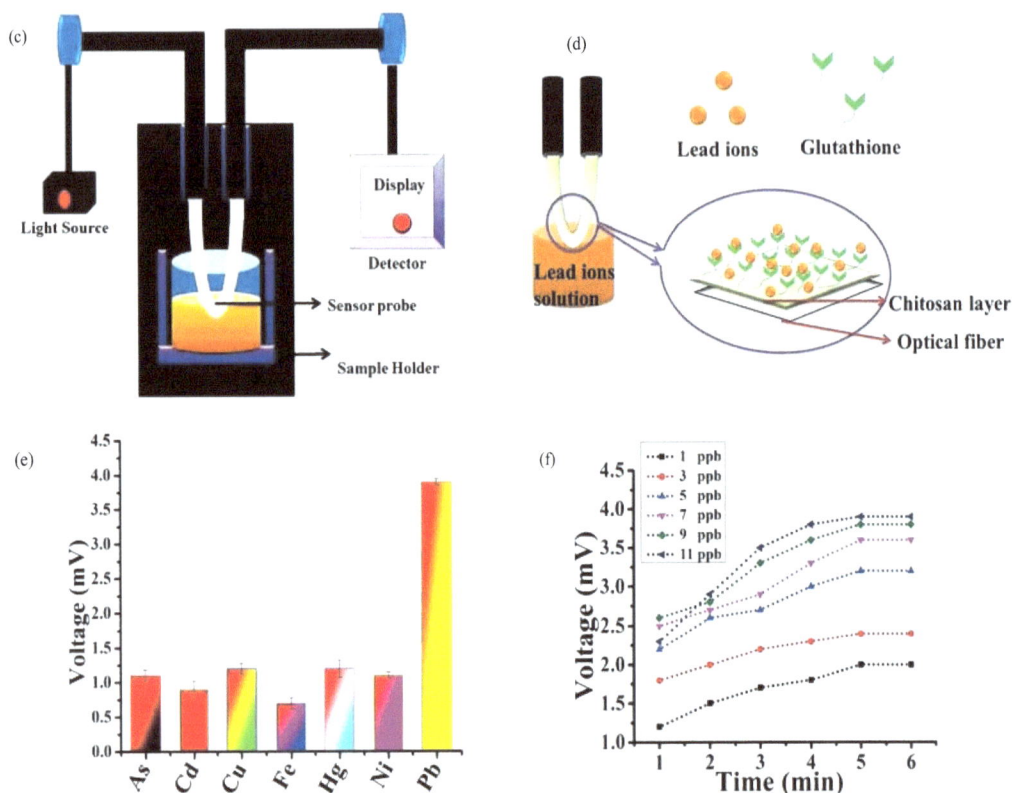

Fig. (3). Optical fiber detection of lead ion (a) preparation of U-shaped optical probe (b) Surface modification of U-shaped probe (c) Optical set up for sensing lead ions sensing unit (d) interaction of lead ions with surface modified sensor probe (e) Output response of sensor unit for different heavy metal ions and (f) Output of sensor system for different concentration lead ions solution [60].

It is apparent that optical fiber sensor unit also offered very low limit of detection and considerable linear range. This implies that optical fiber sensor system may also be an analytical technique for monitoring of heavy metal ion in water samples. Due to rapid detection and ease portability, these sensor systems can be utilized in field application.

Electrochemical Method

The name electrochemical implies electricity which entails current or voltage related with a chemical reaction. Electrochemical method is an analytical method that is used to measure analytes in terms of voltage or current. Electrochemical method deals with charge separation, charge transfer or charge deposition at electrodes. Interaction of chemical compound with the electrodes changes the current or voltage, from which we can measure the concentration of the chemical compound [9, 84]. Of late, application of electrochemical method in various sensing field is remarkably increasing. They are used in environment monitoring, testing food quality, biological sensing application *etc*. Electrochemical method is also found applicable in detection of heavy metal ions in aqueous medium.

In electrochemical method, all necessary reactions take place at the electrodes. The mostly used electrochemical methods for detection of heavy metal ions are potentiometry or voltammetry. In case of potentiometric detection technique, there are only two electrodes, one is working electrode and the other one is reference electrode. In voltammetry detection technique there are three electrodes, reference electrode, working electrode and auxiliary electrode. The commonly used electrodes are gold electrodes, silver electrode and glassy carbon electrode [85]. But due to the low cost, glassy carbon electrode is most commonly used in detection technique. For augmenting the selectivity and sensitivity, working electrode is modified with sensitive materials. The working electrode can be modified by electrochemical deposition, adsorption, electrochemical polymerization and covalent bond formation. Some sensitive materials mostly used for modification of electrodes are gold and silver nanoparticles, magnetic nanoparticles, graphene oxide, chitosan *etc*. For better performance of the sensing system, these sensitive materials are functionalized [86]. In the Fig. (**4a**), schematic of lead ion detection with copper electrode is shown, while Fig. (**4b**) represents the sensor unit. The output response of this sensor unit is shown in Fig. (**5a,b**), for various concentrations of lead ion ranging from 10µM to 25 nM. In the Table **4** provided below, we have shown the recent work based on electrochemical method for heavy metal ion detection. This table has shown the use of different types of electrodes, modified with various types of sensitive materials for detecting various heavy metal ion.

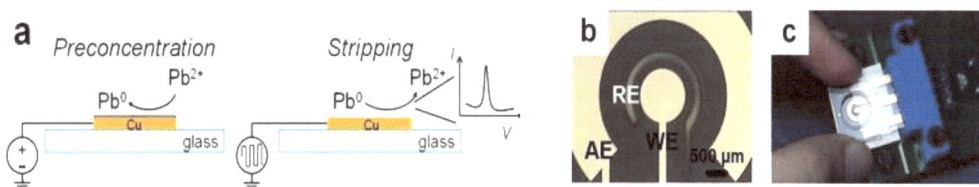

Fig. (4). Schematic of (a) lead ions detection using copper electrodes (b) sensor unit for detection of lead ion [96].

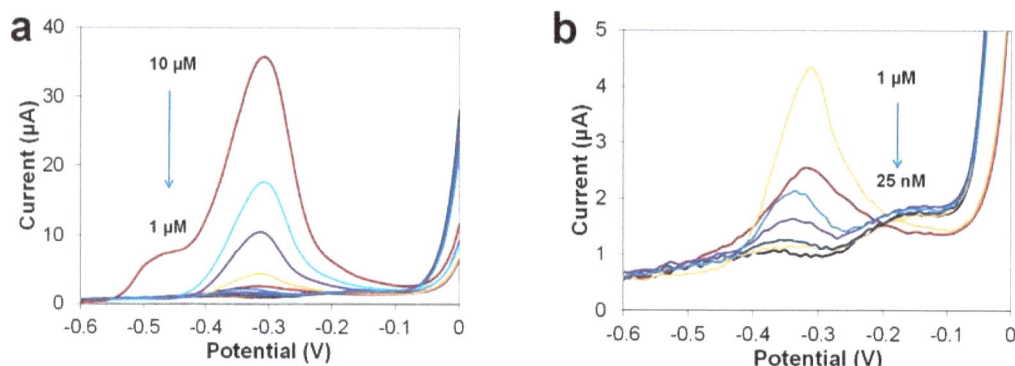

Fig. (5). (a) Current- voltage response for lead ion concentration ranging from 10 μM to 1μM and (b) Current –voltage response of the sensor unit while the concentration ranges from 1μM to 25nM [96].

Table 4. Electrochemical detection of heavy metal ions with various electrodes, different sensitive materials and their linear range and limit of detection.

Electrodes	Sensitive Materials	Heavy Metal Ion	Linear Range	Limit of Detection	References
Glassy carbon electrode	Nano Au-crystal violet	Arsenic	4 - 40 μM	0.20 μM	[87]
Glassy carbon electrode	AgNPs-Graphene Oxide	Arsenic	13.33–375.19 nM	0.24 nM	[88]
Exfoliated graphite	Bismuth	Arsenic	20-100ppb	5ppb	[89]
Carbon fiber ultra-microelectrodes	AuNPs	Arsenic	5-60ppb	0.9ppb	[90]
Gold Electrode	AgNPs	Arsenic	0.05 -0.2 μM	0.0138μM	[91]
Carbon paste electrode	*Eichhornia Crassipes*	Mercury	400-800ppb	195ppb	[92]
ITO-coated glass electrode	Nanostructured bismuth hexagons	Mercury	-	0.74ppb	[93]
Pencil graphite electrode	Deoxyribonucleic acid/poly-L-methionine-AuNPs	Mercury	0.1 aM -0.1 nM	0.004 aM	[94]
Glass carbon electrode	Carboxymethylcellulose-protected AgNPs	Mercury	5 -75 μM	0.19 nM	[95]
Copper Electrode	-	Lead	-	4.4ppb	[96]
Glass carbon electrode	L-Glutamine-ZnO oatmeal-Ag	Lead	5 -6 nM	0.078 nM	[97]
Graphene Paste Electrode	AgNPs	Lead	111.765 - 745.1ppb	12ppb	[98]
Magnetic glassy carbon electrode	Glutathione-magnetic nanoparticles	Lead	0.5-100 ppb	0.182ppb	[99]
Indium tin oxide glass	Co_3O_4 nanosheets	Lead	1-100ppb	0.52	[100]
Screen-printed carbon electrode	graphene/polyaniline/polystyrene nanoporous	Lead	10-500 ppb	3.30ppb	[101]

(Table 4) cont.....

Electrodes	Sensitive Materials	Heavy Metal Ion	Linear Range	Limit of Detection	References
Screen-printed electrode	Single walled carbon nanohorns	Lead	1-60ppb	0.4ppb	[102]
Graphene Paste Electrode	AgNPs	Cadmium	183.96 - 613.2 ppb	17ppb	[98]
Magnetic glassy carbon electrode	Glutathione-magnetic nanoparticles	Cadmium	0.5-10ppb	0.171ppb	[99]
Screen-printed carbon electrode	graphene/polyaniline/polystyrene nanoporous	Cadmium	10-500 ppb	4.43 ppb	[101]
Screen-printed electrode	Single walled carbon nanohorns	Cadmium	1-60 ppb	0.2 ppb	[102]
Screen-printed electrode	Lead film	Nickel	0.6-2.ppb	0.2 ppb	[103]
Glass carbon electrode	N,N'-(ethane-1,2-diyl)bis(3,4-dimethoxybenzenesulfonamide)–Nafion	Nickel	1.0 nM- 1.0 mM	0.78nM	[103]

In electrochemical detection of heavy metal ion, there is a cell where these modified electrodes are kept. When the heavy metal ions solution is placed into the cell, they interact with the electrodes. During interaction electron transfer, oxidation and reduction process take place at the electrodes. Since electrodes are modified with sensitive materials, so heavy metal ions selectively interact with the electrodes. Due to the selective interaction, we get higher voltage- current response for selective heavy metal ion. This implies the selectivity of that metal of the sensor system. Depending on the varying concentration, output response also changes. From the change in output response we can calculate the concentration of heavy metal ion present in that solution.

Mass Spectrometry

It is an analytical method that is usually used for detection of unknown quantity, for knowing the structural properties of materials in samples. Mass spectrometry method is a commonly used scientific method for monitoring of heavy metal ions in water [104, 105]. It is a very precise and accurate method with good sensitivity and selectivity for qualitative and quantitative analysis of unknown samples. Due to high sensitivity and selectivity, mass spectroscopy is used in different fields of science such as environmental science, physical science and biological science [106].

The working principle of a mass spectrometer is the separation of mass to charge ratio according to their abundance. A mass spectrometry has three main components as shown in Fig. (**6**). These are ion source, analyzer and detector unit [106 - 108]. These three components are in a highly vacuum chamber. In the ion source, measured sample is placed. In this ion source chamber, there is high voltage electric field that produces the gaseous form of the sample to be measured. This means that ion source has produced gaseous ions from the sample under investigation. The next unit of the mass spectrometer is the analyzer. In analyzer unit, there are electric and magnetic field of larger magnitudes. These electric and magnetic fields accelerate the gaseous ions. According to the different masses, these gaseous ions are accelerated differently. Heavier ions are slower while lighter ones are faster. It implies that analyzer separates the different gaseous ions according to their masses. The last part of the mass spectrometer is the detector unit. This detector unit collects the abundance of the gaseous ion depending on their masses and produces a mass spectrum. This mass spectrum is the plot of abundance of ions with respect to mass to change ratio. Depending on the mass spectrum plot, we can quantitatively and qualitatively analyze the unknown materials with ease.

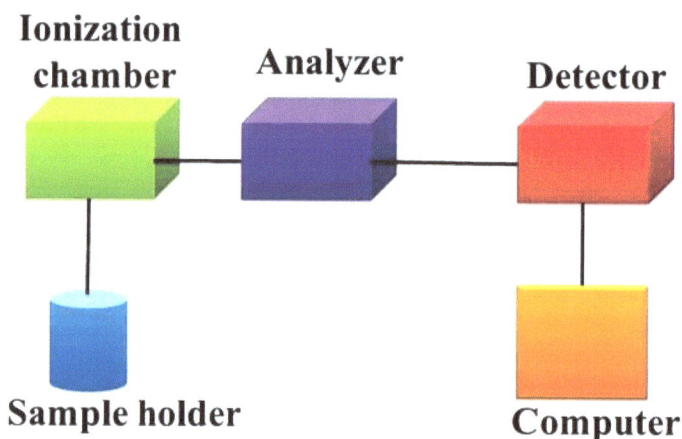

Fig. (6). Schematic of mass spectrometer for detection of heavy metal ions.

In detection of heavy metal ions, mass spectrum produced by mass spectrometer is different for different heavy metal ions. Based on their characteristic spectrum, we can analyze the heavy metal ions present in the solution. Inductive coupled plasma mass spectrometry is an analytical technique for the detection of heavy metal ion. This method provides an excellent detection limit as well as high selectivity and sensitivity. Lu and his research team reported inductively coupled plasma mass spectrometry technique being incorporated with nanomaterials. In their work, they have reported the detection of some toxic heavy metal ions like as lead, cadmium, cobalt, nickel, chromium, copper and silver. Since the concentration of heavy metal ions in environmental samples is very low and they possess complex form, hence Lu and his research team utilized solid phase extraction technique in their work. For selective detection of heavy metal ions in complex solution, silica coated magnetic graphene oxide nanomaterial has been used in their work. This silica coated magnetic graphene oxide nanomaterials were mixed with heavy metal ions solution and then sonicated for few minutes. Afterwards, nanomaterials were separated applying external magnetic field. Towards eluting the heavy metal ions from nanomaterials surface, nitric acid was added and nanomaterials were magnetically separated using external magnetic field. Then, the solid phase of supernatant has been used in inductively coupled plasma mass spectrometry for measuring the concentration of heavy metal ions. In their experiment, pH of all the solution was maintained at 5. This was the optimum pH in their work. This work has provided good sensitivity and linear range along with appreciable limit of detection. Limit of detection for chromium, cobalt, nickel, copper, cadmium, lead and silver was 3.776 ppt, 2.023 ppt, 7.668 ppt, 6.472ppt, 3.794ppt, 3.641 ppt and 13.810 ppt, respectively [112]. In another work, iron detection was demonstrated by Su and his research team using inductively coupled plasma mass spectrometry. In their work, they have used 3D printing technology to build the compact ion exchange chamber. In the conventional inductively coupled plasma mass spectrometry, ion exchange change chamber was packed in commercial available column that requires frits and it limit the flow rate of sample loading. To overcome these problem, Su and his research team have built a compact ion exchange chamber using 3D printing applying non-functionalized acrylate. Then, this chamber is coupled with the mass spectrometer. Here, they have measured the concentration of iron species *i.e.* Fe (II) and Fe (III). The optimum pH of the sample solution is 5. Using this technique, they have found limit of detection for Fe (II) and Fe (III) 1 ppt for each other within the linear range 5-5000 ppt [113]. Similarly, using inductively coupled plasma mass spectrometry, cobalt, nickel, manganese and cadmium detection was reported by Chen and his research team. Instead of using solid phase extraction, they have used dispersive micro-solid phase extraction for quantitative and qualitative analysis of these heavy metal ions. Use of dispersive micro-solid extraction reduces the loss of solvents. For this purpose, they utilized magnetic $ZnFe_2O_4$ nanotube. Through electrospinning and direct annealing method magnetic $ZnFe_2O_3$ nanotubes was synthesized. Then, the solution of the heavy metal ions and the magnetic $ZnFe_2O_3$ solutions were mixed at a pH of 7. pH 7 was the optimum condition of this experiment. Then the mixture solution was sonicated for few minutes to form the dispersion of $ZnFe_2O_3$ nanotubes. Then a strong magnetic field was applied to separate the $ZnFe_2O_3$ from the mixture solution. After this steps nitric acid was added to the $ZnFe_2O_3$ and ultra sonicated the mixture again. Resultant of this process was applied for concentration determination of the target heavy metal ions. Using this dispersive micro-solid phase extraction method incorporating with inductively coupled plasma mass spectrometry, they have obtained the limit of detection for cobalt, nickel, manganese and cadmium 0.10, 1.2, 3.7 and 0.09 pg/ml. This method was also applied for the real environmental sample analysis and results were found to be satisfactory [114]. Heavy metal ions may have different species. For mercury and arsenic that are the most toxic heavy metal ions have various species. In the environment, these metal ions may appear in different forms. Therefore, it is important to determine the species of these heavy metal ions. For this purpose, Fang and his research group have reported a work based on ion pairing reverse chromatography coupled with inductively coupled plasma mass spectrometry. Here, they combine the high performance liquid chromatography with inductively coupled plasma mass spectrometry. They have reported the specification of arsenic and mercury species in rich samples. To specify the arsenic species in rich samples, they have used ion pairing agent like tetrabutylammonium hydroxide (TBAH) and for mercury species, L-cysteine was used. For extraction of mercury and arsenic species from the rich samples, microwave assisted extraction technique was applied. Arsenic species measured in this work were arsenite (As (III)), arsenate (As (V)), monomethylarsonate (MMA) and dimethlyarsinate (DMA). Limit of detection of arsenite, arsenate, monomethylarsonate and dimethlyarsinate was 1.04 ppb, 2.41 ppb, 0.95 ppb and 0.84 ppb, respectively. Similarly, for mercury species, inorganic mercury, methyl mercury and ethyl mercury limit of detection was obtained 0.04 ppb, 0.03 ppb and 0.01 ppb, respectively [115]. In lotus root, inorganic and organic arsenic and mercury species was determined using high performance liquid chromatography coupled with inductively coupled plasma mass spectrometry. TBAH and L-cysteine were used as utilized by Zhang and his research group for selective determination of arsenic and mercury species in lotus root. At optimum pH 6, limit of detection for As (III), DMA, MMA and As (V) were 0.078, 0.081, 0.078 and 0.15 ppb. Limit of detection for mercury species inorganic mercury, MeHg and EtHg were 0.016, 0.027 and 0.032 ppb [116]. Similarly using cloud phase extraction method incorporating with liquid chromatography and inductively coupled plasma mass spectrometry, mercury species determination was done by Chen and his research team [117]. A double cloud point extraction technique in combination with inductively coupled plasma mass spectrometry for detection of various heavy metal ion like as chromium, cadmium, gallium, silver, manganese, iron, indium, copper, nickel, copper, lead and zinc was reported by Liu. For binding of heavy metal ions 8- hydroxyquinoline was used as chelating agent. Under the optical condition like at pH 6.5, temperature 60°C *etc.* linear range of this method for the above heavy metal ions was 1-800ppb. Within this linear range, limit of detection were in between 0.012 and 0.36 ppb [118]. Likewise, mercury species, methylmercury and inorganic mercury detection in water and human hair was reported by Hu and their research team. In this work, they have used magnetic solid phase extraction along with inductively coupled plasma mass spectrometry. Here, they have utilized γ-mercaptopropyltrimethoxysilane (γ-MPTS) functionalized magnetic nanoparticles coated with silica as absorbent for mercury species. Detection of limit for methylmercury and inorganic mercury was 1.6 ppt and 1.9 ppt in the linear range 5-10000 ppt and 10-10000 ppt, respectively [119]. Using the γ-MPTS functionalized silica coated magnetic nanoparticles, Hu reported an another work for detection of mercury species inorganic mercury, methylmercury and phenylmercury in water and fish samples using high performance liquid chromatography coupled with inductively coupled plasma mass spectrometry. Detection limit of these species was found 0.74 ppt, 0.67 ppt and 0.49 ppt within the linear range 5-5000 ppt under the optimum pH 4 [120]. Similarly, Doker and Bosgelmez have reported an inductively coupled plasma mass spectrometry for detection of inorganic mercury and methylmercury in fish sample incorporating with reverse phase liquid chromatography. Limit of detection of this method was 0.2 ng/g and 0.1 ng/g for inorganic mercury and methylmercury [121].

Mesoporous silica functionalized with carboxylic group was applied for the determination of chromium species was reported by Zhu and their research team. In this work, they have used solid phase extraction method combined with inductively coupled plasma mass spectrometry for speciation of chromium ions. Mesoporous silica was prepared through co-condensation method. pH of the solution was maintained at 5 which is the optimum for this experiment. At this pH, chromium (III) interacts with the carboxylic functionalized mesoporous silica but chromium (VI) does not. Therefore, using inductively coupled plasma mass spectrometry with this functionalized materials, chromium (III) was detected. The limit of detection offered by this technique for chromium (III) was 0.02 ppb [122]. Using isotope dilution, chromium (VI) was determined in river sediment using high performance liquid chromatography coupled with inductively coupled plasma mass spectrometry. An alkaline extraction method was utilized for extraction of chromium (VI) from the river sediment. To reduce the oxidation, $MgCl_2$ was added to the extracted chromium sample. Then, chromium isotope 53Cr (III) and 50Cr (VI), were added and applied for the determination of chromium (VI) concentration [123]. Similarly, combining the high performance liquid chromatography with inductively coupled plasma dynamic reaction cell isotope dilution mass spectrometry, total chromium and chromium (VI) was determined by Markiewicz and their research group [124]. Ion chromatography tandem mass spectrometry was applied for the measurement of chromium (VI) in water samples. Without the use of isotope dilution, limit of detection offered by the ion chromatography tandem mass spectrometry was 7ppt. Combination of isotope dilution with ion chromatography tandem mass spectrometry further increased the limit of detection and it was found to be 2 ppt [125].

In Table **5**, we have shown some work for monitoring of heavy metal ions based on mass spectroscopy method.

Table 5. Detection of heavy metal ions through mass spectrometry method, their linear range and limit of detection.

Method	Heavy Metal Ions	Linear Range	LOD	References
ICP-MS	As, Cd, Pb	0.25–3.0 ppb, 0.125–1.5 ppb, 0.25–3.0 ppb	3.1ppb, 0.08ppb, 0.5ppb	[109]
ICP-MS	Mercury	0-1.2 ppb	0.032ppb	[110]
ICP-MS	Cd, Hg, Pb	-	0.002 ppb 0.005ppb 0.001ppb	[111]
ICP-MS	Cr, Ga, Ag, Cd, Mn, Fe, In, Cu, Ni,Co, Pb and Zn	1-800 ppb	0.012 to 0.36 ppb	[118]
HPLC–ICP-MS	Cr (VI), Cr (III)	-	0.1 ppb	[126]
ICP-MS	As(III), As(V)	-	-	[127]
LA-ICP-MS	Mn, Ni, Cu, Zn, As, Se, Cd and Pb	-	-	[128]
ICP-MS	As, Cd and Pb	-	0.04, 0.002, 0.02 µg/g	[129]
ICP-MS	Hg, MeHg	-	0.2 and 0.1 ng/g	[130]
HPLC-ICP-MS	Cr(VI)	-	1.25 µg/kg	[121]
ICP-MS	Cr(VI)	-	0.02 ppb	[123]
ICP-MS	As(III), As(V)	-	2.7 and 3.2 ng/L	[122]
HPLC-ICP-MS	Hg, MeHg, PhHg	-	0.49 to 0.74 ppt	[131]
HPLC-ICP-MS	As(III),DMA,MMA As(V)	-	1.2, 0.96, 0.82 and 0.91 ppt	[120]
IC-MS/MS	Cr (VI)	-	2 ppt	[132]
HPLC–ICP-MS	Cr (III), Cr (VI)	-	0.0068 and 0.0041ng/ml,	[125]
HPLC/ICP-DRC-MS	As(III),As(V),Cr(VI),	5-50 ppb	0.14, 0.062 and 0.15 ppb	[133]
ICP-MS	As(V),Cr(VI), Se(VI)	-	15, 38 and 16 ppt	[134]
ICP-MS	Hg, MeHg, PhHg	-	0.40, 0.49 and 1.4ppt	[135]
ICP-MS	Hg	-	0.75 ppt	[136]
HG-ICP-MS	As	-	2 µg/kg	[137]

FUTURE DIRECTION

Detection of heavy metal ions are very important to sustainable development of society as well as bio-diversity. Application of nanomaterial is a scope for monitoring of heavy metal ions. Towards synthesis and functionalization of nanomaterial, we can use some biological materials that will reduce the toxicity of nanomaterial towards our environment. Another important direction is the combination of optical and electric technique with smart phone sensor that may also help us to make a portable, less time consuming and real time monitoring sensing system.

CONCLUSION

In this chapter, we have discussed different heavy metal ion detection techniques. Towards monitoring heavy metal ions, there exist several approaches such as optical fiber technique, colorimetric technique, mass spectroscopy technique and electrochemical technique. All these techniques possess good sensitivity as well as selectivity for monitoring of heavy metal ions. Among them, colorimetric technique is simple and it is basically detectable through naked eye corresponding to change in color. In general, some colorimetric reagents are used to produce different color for different heavy metal ions. Use of nanomaterial as colorimetric agent is gradually rising because nanomaterial possesses different attribution as from their bulk materials, accompanied by high surface to volume ratio. Application of nanomaterial increases the selectivity and sensitivity and provides reasonable limit of detection and linear range. But, the disadvantage of the colorimetric technique is that some toxic reagents are used as color indicating agent. These toxic reagents again create a problem to environmental health. Electrochemical method is another widely used method for monitoring heavy metal ions. In this method, working electrode is modified with sensitive materials and interaction of heavy metal ion with this electrode is measured through current-voltage distribution. Based on the voltage- current response, heavy metal ions are monitored in samples. Although electrochemical method provides good sensitivity and selectivity towards detection of heavy metal ions, the problem of this method is the requirement of gold electrodes for better performance. Utilization of gold electrodes makes the sensor system costly. Likewise, mass spectrometry method is a powerful technique for heavy metal ion detection. It provides precise and accurate measurement. Requirement of expensive experimental setup hinders field application. Apart from that, it requires large amount of sample solution and sample treatment process is time consuming. Another most widely used method is optical fiber technique, where an optical fiber is used as sensor unit. Using optical fiber sensor unit, many research groups have detected different heavy metal ions with reliable sensitivity. Utilization of light emitting diode as light source and photo diode as detector makes this sensor unit low cost and compact. Due to compact nature, optical fiber sensor system can be applied for field application.

CONSENT FOR PUBLICATION

Not applicable.

CONFLICT OF INTEREST

The authors confirm that this chapter contents have no conflict of interest.

ACKNOWLEDGEMENT

Author acknowledges Bentham Science Publisher for inviting to write this book chapter.

REFERENCES

[1]   Turdean, G.L. Design and development of biosensors for the detection of heavy metal toxicity. *Int. J. Electrochem.,* **2011**, *2011*, 15.
[http://dx.doi.org/10.4061/2011/343125]

[2]   Ullah, N.; Mansha, M.; Khan, I.; Qurashi, A. Nanomaterial-based optical chemical sensors for the detection of heavy metals in water: Recent advances and challenges. *Trends Analyt. Chem.,* **2018**, *100*, 155-166.
[http://dx.doi.org/10.1016/j.trac.2018.01.002]

[3]   Anjum, M.; Miandad, R.; Waqas, M.; Gehany, F.; Barakat, M.A. Remediation of wastewater using various nano-materials. *Arab. J. Chem.,* **2016**, *12*(8)
[http://dx.doi.org/10.1016/j.arabjc.2016.10.004]

[4]   Wu, Y.; Pang, H.; Liu, Y.; Wang, X.; Yu, S.; Fu, D.; Wang, X. Environmental remediation of heavy metal ions by novel-nanomaterials: a review. *Environ. Pollut.,* **2018**.
[PMID: 30605816]

[5]   Mandal, B.K.; Suzuki, K.T. Arsenic round the world: a review. *Talanta,* **2002**, *58*(1), 201-235.
[http://dx.doi.org/10.1016/S0039-9140(02)00268-0] [PMID: 18968746]

[6]   Feng, L.; Cao, M.; Ma, X.; Zhu, Y.; Hu, C. Superparamagnetic high-surface-area Fe3O4 nanoparticles as adsorbents for arsenic removal. *J. Hazard. Mater.,* **2012**, *217-218*, 439-446.
[http://dx.doi.org/10.1016/j.jhazmat.2012.03.073] [PMID: 22494901]

[7]   Park, G.J.; You, G.R.; Choi, Y.W.; Kim, C. A naked-eye chemosensor for simultaneous detection of iron and copper ions and its copper complex for colorimetric/fluorescent sensing of cyanide. *Sens. Actuators B Chem.,* **2016**, *229*, 257-271.
[http://dx.doi.org/10.1016/j.snb.2016.01.133]

[8]   Narayanan, K.B.; Park, H.H. Colorimetric detection of manganese(II) ions using gold/dopa nanoparticles. *Spectrochim. Acta A Mol. Biomol. Spectrosc.,* **2014**, *131*, 132-137.
[http://dx.doi.org/10.1016/j.saa.2014.04.081] [PMID: 24825666]

[9]   Bansod, B.; Kumar, T.; Thakur, R.; Rana, S.; Singh, I. A review on various electrochemical techniques for heavy metal ions detection with different sensing platforms. *Biosens. Bioelectron.,* **2017**, *94*, 443-455.

[http://dx.doi.org/10.1016/j.bios.2017.03.031] [PMID: 28340464]

[10]   Hegazi, H.A. Removal of heavy metals from wastewater using agricultural and industrial wastes as adsorbents. *HBRC Journal,* **2013**, *9*(3), 276-282.
[http://dx.doi.org/10.1016/j.hbrcj.2013.08.004]

[11]   Abu-El-Halawa, R.; Zabin, S.A. Removal efficiency of Pb, Cd, Cu and Zn from polluted water using dithiocarbamate ligands. *Journal of Taibah University for Science,* **2017**, *11*(1), 57-65.
[http://dx.doi.org/10.1016/j.jtusci.2015.07.002]

[12]   Jan, A.T.; Azam, M.; Siddiqui, K.; Ali, A.; Choi, I.; Haq, Q.M. Heavy metals and human health: mechanistic insight into toxicity and counter defense system of antioxidants. *Int. J. Mol. Sci.,* **2015**, *16*(12), 29592-29630.
[http://dx.doi.org/10.3390/ijms161226183] [PMID: 26690422]

[13]   Miao, P.; Tang, Y.; Wang, L. DNA modified $Fe_3O_4@$ Au magnetic nanoparticles as selective probes for simultaneous detection of heavy metal ions. *ACS Appl. Mater. Interfaces,* **2017**, *9*(4), 3940-3947.
[http://dx.doi.org/10.1021/acsami.6b14247] [PMID: 28079364]

[14]   Li, M.; Gou, H.; Al-Ogaidi, I.; Wu, N. Nanostructured sensors for detection of heavy metals: a review. *ACS Sustain. Chem.& Eng.,* **2013**, *1*, 713-723.
[http://dx.doi.org/10.1021/sc400019a]

[15]   Li, F.; Wang, J.; Lai, Y.; Wu, C.; Sun, S.; He, Y.; Ma, H. Ultrasensitive and selective detection of copper (II) and mercury (II) ions by dye-coded silver nanoparticle-based SERS probes. *Biosens. Bioelectron.,* **2013**, *39*(1), 82-87.
[http://dx.doi.org/10.1016/j.bios.2012.06.050] [PMID: 22840330]

[16]   Melamed, D. Monitoring arsenic in the environment: a review of science and technologies with the potential for field measurements. *Anal. Chim. Acta,* **2005**, *532*(1), 1-13.
[http://dx.doi.org/10.1016/j.aca.2004.10.047]

[17]   Wang, A.J.; Guo, H.; Zhang, M.; Zhou, D.L.; Wang, R.Z.; Feng, J.J. Sensitive and selective colorimetric detection of cadmium (II) using gold nanoparticles modified with 4-amino-3-hydrazi-o-5-mercapto-1, 2, 4-triazole. *Mikrochim. Acta,* **2013**, *180*(11-12), 1051-1057.
[http://dx.doi.org/10.1007/s00604-013-1030-7]

[18]   Taghdisi, S.M.; Danesh, N.M.; Lavaee, P.; Ramezani, M.; Abnous, K. An aptasensor for selective, sensitive and fast detection of lead(II) based on polyethyleneimine and gold nanoparticles. *Environ. Toxicol. Pharmacol.,* **2015**, *39*(3), 1206-1211.
[http://dx.doi.org/10.1016/j.etap.2015.04.013] [PMID: 25989533]

[19]   Li, J.J.; Hou, C.J.; Huo, D.Q.; Shen, C.H.; Luo, X.G.; Fa, H.B.; Zhou, J. Detection of trace nickel ions with a colorimetric sensor based on indicator displacement mechanism. *Sens. Actuators B Chem.,* **2017**, *241*, 1294-1302.
[http://dx.doi.org/10.1016/j.snb.2016.09.191]

[20]   Kanagaraj, R.; Nam, Y.S.; Pai, S.J.; Han, S.S.; Lee, K.B. Highly selective and sensitive detection of Cr6+ ions using size-specific label-free gold nanoparticles. *Sens. Actuators B Chem.,* **2017**, *251*, 683-691.
[http://dx.doi.org/10.1016/j.snb.2017.05.089]

[21]   Tan, F.; Cong, L.; Jiang, X.; Wang, Y.; Quan, X.; Chen, J.; Mulchandani, A. Highly sensitive detection of Cr (VI) by reduced graphene oxide chemiresistor and 1, 4-dithiothreitol functionalized Au nanoparticles. *Sens. Actuators B Chem.,* **2017**, *247*, 265-272.
[http://dx.doi.org/10.1016/j.snb.2017.02.163]

[22]   Wang, X.; Li, F.; Cai, Z.; Liu, K.; Li, J.; Zhang, B.; He, J. Sensitive colorimetric assay for uric acid and glucose detection based on multilayer-modified paper with smartphone as signal readout. *Anal. Bioanal. Chem.,* **2018**, *410*(10), 2647-2655.
[http://dx.doi.org/10.1007/s00216-018-0939-4] [PMID: 29455281]

[23]   Murray, E.; Nesterenko, E.P.; McCaul, M.; Morrin, A.; Diamond, D.; Moore, B. A colorimetric method for use within portable test kits for nitrate determination in various water matrices. *Anal. Methods,* **2017**, *9*(4), 680-687.
[http://dx.doi.org/10.1039/C6AY03190K]

[24]   Sharma, H.; Kaur, N.; Singh, A.; Kuwar, A.; Singh, N. Optical chemosensors for water sample analysis. *J. Mater. Chem. C Mater. Opt. Electron. Devices,* **2016**, *4*(23), 5154-5194.
[http://dx.doi.org/10.1039/C6TC00605A]

[25]   Priyadarshini, E.; Pradhan, N. Gold nanoparticles as efficient sensors in colorimetric detection of toxic metal ions: a review. *Sens. Actuators B Chem.,* **2017**, *238*, 888-902.
[http://dx.doi.org/10.1016/j.snb.2016.06.081]

[26]   Feng, B.; Zhu, R.; Xu, S.; Chen, Y.; Di, J. A sensitive LSPR sensor based on glutathione-functionalized gold nanoparticles on a substrate for the detection of Pb 2+ ions. *RSC Advances,* **2018**, *8*(8), 4049-4056.
[http://dx.doi.org/10.1039/C7RA13127E]

[27]   Ratnarathorn, N.; Chailapakul, O.; Dungchai, W. Highly sensitive colorimetric detection of lead using maleic acid functionalized gold nanoparticles. *Talanta,* **2015**, *132*, 613-618.
[http://dx.doi.org/10.1016/j.talanta.2014.10.024] [PMID: 25476352]

[28]   Huang, K.W.; Yu, C.J.; Tseng, W.L. Sensitivity enhancement in the colorimetric detection of lead(II) ion using gallic acid-capped gold nanoparticles: improving size distribution and minimizing interparticle repulsion. *Biosens. Bioelectron.,* **2010**, *25*(5), 984-989.
[http://dx.doi.org/10.1016/j.bios.2009.09.006] [PMID: 19782557]

[29]   Qi, L.; Shang, Y.; Wu, F. Colorimetric detection of lead (II) based on silver nanoparticles capped with iminodiacetic acid. *Mikrochim. Acta,* **2012**, *178*(1-2), 221-227.
[http://dx.doi.org/10.1007/s00604-012-0832-3]

[30]   Noh, K.C.; Nam, Y.S.; Lee, H.J.; Lee, K.B. A colorimetric probe to determine Pb(2+) using functionalized silver nanoparticles. *Analyst (Lond.),* **2015**, *140*(24), 8209-8216.
[http://dx.doi.org/10.1039/C5AN01601K] [PMID: 26555436]

[31]   Boruah, B.S.; Daimari, N.K.; Biswas, R. Functionalized silver nanoparticles as an effective medium towards trace determination of arsenic (III) in aqueous solution. *Results in Physics,* **2019**, *12*, 2061-2065.
[http://dx.doi.org/10.1016/j.rinp.2019.02.044]

[32]   Divsar, F.; Habibzadeh, K.; Shariati, S.; Shahriarinour, M. Aptamer conjugated silver nanoparticles for the colorimetric detection of arsenic ions using response surface methodology. *Anal. Methods,* **2015**, *7*(11), 4568-4576.
[http://dx.doi.org/10.1039/C4AY02914C]

[33]   Boruah, B.S.; Biswas, R. Selective detection of arsenic (III) based on colorimetric approach in aqueous medium using functionalized gold nanoparticles unit. *Mater. Res. Express,* **2018**, *5*(1)015059
[http://dx.doi.org/10.1088/2053-1591/aaa661]

[34]   Boruah, B.S.; Biswas, R.; Deb, P. A green colorimetric approach towards detection of arsenic (III): A pervasive environmental pollutant. *Opt. Laser Technol.,* **2019**, *111*, 825-829.
[http://dx.doi.org/10.1016/j.optlastec.2018.09.023]

[35]   Domínguez-González, R.; González Varela, L.; Bermejo-Barrera, P. Functionalized gold nanoparticles for the detection of arsenic in water. *Talanta,* **2014**, *118*, 262-269.
[http://dx.doi.org/10.1016/j.talanta.2013.10.029] [PMID: 24274297]

[36]   Boruah, B.S.; Daimari, N.K.; Biswas, R. Mangifera indica leaf extract mediated gold nanoparticles: a novel platform for sensing of As (III). *IEEE Sensors Letters,* **2019**, *3*(3), 1-3.
[http://dx.doi.org/10.1109/LSENS.2019.2894419]

[37]   Sener, G.; Uzun, L.; Denizli, A. Lysine-promoted colorimetric response of gold nanoparticles: a

simple assay for ultrasensitive mercury(II) detection. *Anal. Chem.,* **2014**, *86*(1), 514-520.
[http://dx.doi.org/10.1021/ac403447a] [PMID: 24364626]

[38]   Chen, G.H.; Chen, W.Y.; Yen, Y.C.; Wang, C.W.; Chang, H.T.; Chen, C.F. Detection of mercury(II) ions using colorimetric gold nanoparticles on paper-based analytical devices. *Anal. Chem.,* **2014**, *86*(14), 6843-6849.
[http://dx.doi.org/10.1021/ac5008688] [PMID: 24932699]

[39]   Khan, U.; Niaz, A.; Shah, A.; Zaman, M.I.; Zia, M.A.; Iftikhar, F.J.; Shah, A.H. Thiamine-functionalized silver nanoparticles for the highly selective and sensitive colorimetric detection of Hg 2+ ions. *New J. Chem.,* **2018**, *42*(1), 528-534.
[http://dx.doi.org/10.1039/C7NJ03382F]

[40]   Lee, J.S.; Han, M.S.; Mirkin, C.A. Colorimetric detection of mercuric ion (Hg2+) in aqueous media using DNA-functionalized gold nanoparticles. *Angew. Chem. Int. Ed. Engl.,* **2007**, *46*(22), 4093-4096.
[http://dx.doi.org/10.1002/anie.200700269] [PMID: 17461429]

[41]   Maity, D.; Kumar, A.; Gunupuru, R.; Paul, P. Colorimetric detection of mercury (II) in aqueous media with high selectivity using calixarene functionalized gold nanoparticles. *Colloids Surf. A Physicochem. Eng. Asp.,* **2014**, *455*, 122-128.
[http://dx.doi.org/10.1016/j.colsurfa.2014.04.047]

[42]   Su, D.; Yang, X.; Xia, Q.; Chai, F.; Wang, C.; Qu, F. Colorimetric detection of Hg 2+ using thioctic acid functionalized gold nanoparticles. *RSC Advances,* **2013**, *3*(46), 24618-24624.
[http://dx.doi.org/10.1039/c3ra43276a]

[43]   Mehta, V.N.; Basu, H.; Singhal, R.K.; Kailasa, S.K. Simple and sensitive colorimetric sensing of Cd2+ ion using chitosan dithiocarbamate functionalized gold nanoparticles as a probe. *Sens. Actuators B Chem.,* **2015**, *220*, 850-858.
[http://dx.doi.org/10.1016/j.snb.2015.05.105]

[44]   Sung, Y.M.; Wu, S.P. Colorimetric detection of Cd (II) ions based on di-(1H-pyrrol-2-yl) methanethione functionalized gold nanoparticles. *Sens. Actuators B Chem.,* **2014**, *201*, 86-91.
[http://dx.doi.org/10.1016/j.snb.2014.04.069]

[45]   Xue, Y.; Zhao, H.; Wu, Z.; Li, X.; He, Y.; Yuan, Z. Colorimetric detection of Cd2+ using gold nanoparticles cofunctionalized with 6-mercaptonicotinic acid and L-cysteine. *Analyst (Lond.),* **2011**, *136*(18), 3725-3730.
[http://dx.doi.org/10.1039/c1an15238f] [PMID: 21804959]

[46]   Jin, W.; Huang, P.; Wu, F.; Ma, L.H. Ultrasensitive colorimetric assay of cadmium ion based on silver nanoparticles functionalized with 5-sulfosalicylic acid for wide practical applications. *Analyst (Lond.),* **2015**, *140*(10), 3507-3513.
[http://dx.doi.org/10.1039/C5AN00230C] [PMID: 25831211]

[47]   Du, J.; Hu, X.; Zhang, G.; Wu, X.; Gong, D. Colorimetric detection of cadmium in water using L-cysteine Functionalized gold–silver nanoparticles. *Anal. Lett.,* **2018**, *51*(18), 2906-2919.
[http://dx.doi.org/10.1080/00032719.2018.1455103]

[48]   Chen, N.; Zhang, Y.; Liu, H.; Ruan, H.; Dong, C.; Shen, Z.; Wu, A. A supersensitive probe for rapid colorimetric detection of nickel ion based on a sensing mechanism of anti-etching. *ACS Sustain. Chem.& Eng.,* **2016**, *4*(12), 6509-6516.
[http://dx.doi.org/10.1021/acssuschemeng.6b01326]

[49]   Parnsubsakul, A.; Oaew, S.; Surareungchai, W. Zwitterionic peptide-capped gold nanoparticles for colorimetric detection of Ni²⁺. *Nanoscale,* **2018**, *10*(12), 5466-5473.
[http://dx.doi.org/10.1039/C7NR07998B] [PMID: 29445795]

[50]   Feng, J.; Jin, W.; Huang, P.; Wu, F. Highly selective colorimetric detection of Ni 2+ using silver nanoparticles cofunctionalized with adenosine monophosphate and sodium dodecyl sulfonate. *J. Nanopart. Res.,* **2017**, *19*(9), 306.
[http://dx.doi.org/10.1007/s11051-017-3998-0]

[51]   Kiatkumjorn, T.; Rattanarat, P.; Siangproh, W.; Chailapakul, O.; Praphairaksit, N. Glutathione and L-cysteine modified silver nanoplates-based colorimetric assay for a simple, fast, sensitive and selective determination of nickel. *Talanta,* **2014**, *128*, 215-220.
[http://dx.doi.org/10.1016/j.talanta.2014.04.085] [PMID: 25059151]

[52]   Shang, Y.; Wu, F.; Qi, L. Highly selective colorimetric assay for nickel ion using N-acetyl-l-cysteine-functionalized silver nanoparticles. *J. Nanopart. Res.,* **2012**, *14*(10), 1169.
[http://dx.doi.org/10.1007/s11051-012-1169-x]

[53]   Gupta, B.D.; Dodeja, H.; Tomar, A.K. Fibre-optic evanescent field absorption sensor based on a U-shaped probe. *Opt. Quantum Electron.,* **1996**, *28*(11), 1629-1639.
[http://dx.doi.org/10.1007/BF00331053]

[54]   Satija, J.; Punjabi, N.S.; Sai, V.V.R.; Mukherji, S. Optimal design for U-bent fiber-optic LSPR sensor probes. *Plasmonics,* **2014**, *9*(2), 251-260.
[http://dx.doi.org/10.1007/s11468-013-9618-7]

[55]   Zhao, Y.; Li, X.G.; Zhou, X.; Zhang, Y.N. Review on the graphene based optical fiber chemical and biological sensors. *Sens. Actuators B Chem.,* **2016**, *231*, 324-340.
[http://dx.doi.org/10.1016/j.snb.2016.03.026]

[56]   Sai, V.V.R.; Kundu, T.; Mukherji, S. Novel U-bent fiber optic probe for localized surface plasmon resonance based biosensor. *Biosens. Bioelectron.,* **2009**, *24*(9), 2804-2809.
[http://dx.doi.org/10.1016/j.bios.2009.02.007] [PMID: 19285853]

[57]   Homola, J. Surface plasmon resonance sensors for detection of chemical and biological species. *Chem. Rev.,* **2008**, *108*(2), 462-493.
[http://dx.doi.org/10.1021/cr068107d] [PMID: 18229953]

[58]   Unser, S.; Bruzas, I.; He, J.; Sagle, L. Localized surface plasmon resonance biosensing: current challenges and approaches. *Sensors (Basel),* **2015**, *15*(7), 15684-15716.
[http://dx.doi.org/10.3390/s150715684] [PMID: 26147727]

[59]   Ortega-Mendoza, J.G.; Padilla-Vivanco, A.; Toxqui-Quitl, C.; Zaca-Morán, P.; Villegas-Hernández, D.; Chávez, F. Optical fiber sensor based on localized surface plasmon resonance using silver nanoparticles photodeposited on the optical fiber end. *Sensors (Basel),* **2014**, *14*(10), 18701-18710.
[http://dx.doi.org/10.3390/s141018701] [PMID: 25302813]

[60]   Boruah, B.S.; Biswas, R. An optical fiber based surface plasmon resonance technique for sensing of lead ions: A toxic water pollutant. *Opt. Fiber Technol.,* **2018**, *46*, 152-156.
[http://dx.doi.org/10.1016/j.yofte.2018.10.007]

[61]   Boruah, B.S.; Biswas, R. Localized surface plasmon resonance based U-shaped optical fiber probe for the detection of Pb2+ in aqueous medium. *Sens. Actuators B Chem.,* **2018**, *276*, 89-94.
[http://dx.doi.org/10.1016/j.snb.2018.08.086]

[62]   Shukla, G.M.; Punjabi, N.; Kundu, T.; Mukherji, S. Optimization of Plasmonic U-shaped Optical Fiber Sensor for Mercury Ions Detection using Glucose Capped Silver Nanoparticles. *IEEE Sens. J.,* **2019**, *19*, 3224-3231.
[http://dx.doi.org/10.1109/JSEN.2019.2893270]

[63]   Halkare, P.; Punjabi, N.; Wangchuk, J.; Nair, A.; Kondabagil, K.; Mukherji, S. Bacteria functionalized gold nanoparticle matrix based fiber-optic sensor for monitoring heavy metal pollution in water. *Sens. Actuators B Chem.,* **2019**, *281*, 643-651.
[http://dx.doi.org/10.1016/j.snb.2018.10.119]

[64]   Long, F.; Zhu, A.; Shi, H.; Wang, H.; Liu, J. Rapid on-site/in-situ detection of heavy metal ions in environmental water using a structure-switching DNA optical biosensor. *Sci. Rep.,* **2013**, *3*, 2308.
[http://dx.doi.org/10.1038/srep02308] [PMID: 23892693]

[65]   Benounis, M.; Jaffrezic-Renault, N.; Halouani, H.; Lamartine, R.; Dumazet-Bonnamour, I. Detection of heavy metals by an optical fiber sensor with a sensitive cladding including a new chromogenic calix

[4] arene molecule. *Mater. Sci. Eng. C,* **2006**, *26*(2-3), 364-368.
[http://dx.doi.org/10.1016/j.msec.2005.10.055]

[66]   Verma, R.; Gupta, B.D. Detection of heavy metal ions in contaminated water by surface plasmon resonance based optical fibre sensor using conducting polymer and chitosan. *Food Chem.,* **2015**, *166*, 568-575.
[http://dx.doi.org/10.1016/j.foodchem.2014.06.045] [PMID: 25053095]

[67]   Laxmeshwar, L.S.; Jadhav, M.S.; Akki, J.F.; Raikar, P.; Raikar, U.S. Elemental analysis of wastewater effluent using highly sensitive fiber Bragg grating sensor. *Opt. Laser Technol.,* **2018**, *105*, 45-51.
[http://dx.doi.org/10.1016/j.optlastec.2018.02.043]

[68]   Lin, T.J.; Chung, M.F. Detection of cadmium by a fiber-optic biosensor based on localized surface plasmon resonance. *Biosens. Bioelectron.,* **2009**, *24*(5), 1213-1218.
[http://dx.doi.org/10.1016/j.bios.2008.07.013] [PMID: 18718753]

[69]   Tagad, C.K.; Kulkarni, A.; Aiyer, R.C.; Patil, D.; Sabharwal, S.G. A miniaturized optical biosensor for the detection of Hg2+ based on acid phosphatase inhibition. *Optik (Stuttg.),* **2016**, *127*(20), 8807-8811.
[http://dx.doi.org/10.1016/j.ijleo.2016.06.123]

[70]   Chandra, S.; Dhawangale, A.; Mukherji, S. Hand-held optical sensor using denatured antibody coated electro-active polymer for ultra-trace detection of copper in blood serum and environmental samples. *Biosens. Bioelectron.,* **2018**, *110*, 38-43.
[http://dx.doi.org/10.1016/j.bios.2018.03.040] [PMID: 29587192]

[71]   Ibrahim, S.A.; Ridzwan, A.H.; Mansoor, A.; Dambul, K.D. Tapered optical fibre coated with chitosan for lead (II) ion sensing. *Electron. Lett.,* **2016**, *52*(12), 1049-1050.
[http://dx.doi.org/10.1049/el.2016.0762]

[72]   Aliberti, A.; Vaiano, P.; Caporale, A.; Consales, M.; Ruvo, M.; Cusano, A. Fluorescent chemosensors for Hg2+ detection in aqueous environment. *Sens. Actuators B Chem.,* **2017**, *247*, 727-735.
[http://dx.doi.org/10.1016/j.snb.2017.03.026]

[73]   Long, F.; Gao, C.; Shi, H.C.; He, M.; Zhu, A.N.; Klibanov, A.M.; Gu, A.Z. Reusable evanescent wave DNA biosensor for rapid, highly sensitive, and selective detection of mercury ions. *Biosens. Bioelectron.,* **2011**, *26*(10), 4018-4023.
[http://dx.doi.org/10.1016/j.bios.2011.03.022] [PMID: 21550227]

[74]   Raghunandhan, R.; Chen, L.H.; Long, H.Y.; Leam, L.L.; So, P.L.; Ning, X.; Chan, C.C. Chitosan/PAA based fiber-optic interferometric sensor for heavy metal ions detection. *Sens. Actuators B Chem.,* **2016**, *233*, 31-38.
[http://dx.doi.org/10.1016/j.snb.2016.04.020]

[75]   Ivask, A.; Green, T.; Polyak, B.; Mor, A.; Kahru, A.; Virta, M.; Marks, R. Fibre-optic bacterial biosensors and their application for the analysis of bioavailable Hg and As in soils and sediments from Aznalcollar mining area in Spain. *Biosens. Bioelectron.,* **2007**, *22*(7), 1396-1402.
[http://dx.doi.org/10.1016/j.bios.2006.06.019] [PMID: 16889954]

[76]   Yanaz, Z.; Filik, H.; Apak, R. Development of an optical fibre reflectance sensor for lead detection based on immobilised arsenazo III. *Sens. Actuators B Chem.,* **2010**, *147*(1), 15-22.
[http://dx.doi.org/10.1016/j.snb.2009.12.024]

[77]   Pérez-Hernández, J.; Albero, J.; Llobet, E.; Correig, X.; Matías, I.R.; Arregui, F.J.; Palomares, E. Mercury optical fibre probe based on a modified cladding of sensitised Al2O3 nano-particles. *Sens. Actuators B Chem.,* **2009**, *143*(1), 103-110.
[http://dx.doi.org/10.1016/j.snb.2009.08.051]

[78]   Yeh, T.C.; Tien, P.; Chau, L.K. Fiber-optic evanescent-wave absorption copper (II) sensor based on sol-gel-derived organofunctionalized silica cladding. *Appl. Spectrosc.,* **2001**, *55*(10), 1320-1326.
[http://dx.doi.org/10.1366/0003702011953676]

[79]   Tan, S.Y.; Lee, S.C.; Okazaki, T.; Kuramitz, H.; Abd-Rahman, F. Detection of mercury (II) ions in

water by polyelectrolyte–gold nanoparticles coated long period fiber grating sensor. *Opt. Commun.,* **2018**, *419*, 18-24.
[http://dx.doi.org/10.1016/j.optcom.2018.02.069]

[80]   Yap, S.H.K.; Chien, Y.H.; Tan, R.; Bin Shaik Alauddin, A.R.; Ji, W.B.; Tjin, S.C.; Yong, K.T. An Advanced Hand-Held Microfiber-Based Sensor for Ultrasensitive Lead Ion Detection. *ACS Sens.,* **2018**, *3*(12), 2506-2512.
[http://dx.doi.org/10.1021/acssensors.8b01031] [PMID: 30421612]

[81]   Tran, V.T.; Tran, N.H.T.; Nguyen, T.T.; Yoon, W.J.; Ju, H. Liquid Cladding Mediated Optical Fiber Sensors for Copper Ion Detection. *Micromachines (Basel),* **2018**, *9*(9), 471.
[http://dx.doi.org/10.3390/mi9090471] [PMID: 30424404]

[82]   Lin, T.J.; Chung, M.F. Using monoclonal antibody to determine lead ions with a localized surface plasmon resonance fiber-optic biosensor. *Sensors (Basel),* **2008**, *8*(1), 582-593.
[http://dx.doi.org/10.3390/s8010582] [PMID: 27879723]

[83]   Raj, D.R.; Prasanth, S.; Vineeshkumar, T.V.; Sudarsanakumar, C. Surface plasmon resonance based fiber optic sensor for mercury detection using gold nanoparticles PVA hybrid. *Opt. Commun.,* **2016**, *367*, 102-107.
[http://dx.doi.org/10.1016/j.optcom.2016.01.027]

[84]   Dai, X.; Wu, S.; Li, S. Progress on electrochemical sensors for the determination of heavy metal ions from contaminated water. *Journal of the Chinese Advanced Materials Society,* **2018**, *6*(2), 91-111.
[http://dx.doi.org/10.1080/22243682.2018.1425904]

[85]   March, G.; Nguyen, T.D.; Piro, B. Modified electrodes used for electrochemical detection of metal ions in environmental analysis. *Biosensors (Basel),* **2015**, *5*(2), 241-275.
[http://dx.doi.org/10.3390/bios5020241] [PMID: 25938789]

[86]   Kempahanumakkagari, S.; Deep, A.; Kim, K.H.; Kumar Kailasa, S.; Yoon, H.O. Nanomaterial-based electrochemical sensors for arsenic - A review. *Biosens. Bioelectron.,* **2017**, *95*, 106-116.
[http://dx.doi.org/10.1016/j.bios.2017.04.013] [PMID: 28431363]

[87]   Rajkumar, M.; Thiagarajan, S.; Chen, S.M. Electrochemical detection of arsenic in various water samples. *Int. J. Electrochem. Sci.,* **2011**, *6*(8), 3164-3177.

[88]   Dar, R.A.; Khare, N.G.; Cole, D.P.; Karna, S.P.; Srivastava, A.K. Green synthesis of a silver nanoparticle–graphene oxide composite and its application for As (III) detection. *RSC Advances,* **2014**, *4*(28), 14432-14440.
[http://dx.doi.org/10.1039/C4RA00934G]

[89]   Ndlovu, T.; Mamba, B.B.; Sampath, S.; Krause, R.W.; Arotiba, O.A. Voltammetric detection of arsenic on a bismuth modified exfoliated graphite electrode. *Electrochim. Acta,* **2014**, *128*, 48-53.
[http://dx.doi.org/10.1016/j.electacta.2013.08.084]

[90]   Carrera, P.; Espinoza-Montero, P.J.; Fernández, L.; Romero, H.; Alvarado, J. Electrochemical determination of arsenic in natural waters using carbon fiber ultra-microelectrodes modified with gold nanoparticles. *Talanta,* **2017**, *166*, 198-206.
[http://dx.doi.org/10.1016/j.talanta.2017.01.056] [PMID: 28213223]

[91]   Sonkoue, B.M.; Tchekwagep, P.M.S.; Nanseu Njiki, C.P.; Ngameni, E. Electrochemical Determination of Arsenic Using Silver Nanoparticles. *Electroanalysis,* **2018**, *30*(11), 2738-2743.
[http://dx.doi.org/10.1002/elan.201800520]

[92]   Rajawat, D.S.; Srivastava, S.; Satsangee, S.P. Electrochemical determination of mercury at trace levels using eichhornia crassipes modified carbon paste electrode. *Int. J. Electrochem. Sci.,* **2012**, *7*, 11456-11469.

[93]   Gupta, S.; Singh, R.; Anoop, M.D.; Kulshrestha, V.; Srivastava, D.N.; Ray, K.; Kumar, M. Electrochemical sensor for detection of mercury (II) ions in water using nanostructured bismuth hexagons. *Appl. Phys., A Mater. Sci. Process.,* **2018**, *124*(11), 737.

[http://dx.doi.org/10.1007/s00339-018-2161-9]

[94]   Akbari Hasanjani, H.R.; Zarei, K. An electrochemical sensor for attomolar determination of mercury(II) using DNA/poly-L-methionine-gold nanoparticles/pencil graphite electrode. *Biosens. Bioelectron.,* **2019**, *128,* 1-8.
[http://dx.doi.org/10.1016/j.bios.2018.12.039] [PMID: 30616212]

[95]   Meenakshi, S.; Devi, S.; Pandian, K.; Chitra, K.; Tharmaraj, P. Aniline-mediated synthesis of carboxymethyl cellulose protected silver nanoparticles modified electrode for the differential pulse anodic stripping voltammetry detection of mercury at trace level. *Ionics,* **2019**, •••, 1-11.
[http://dx.doi.org/10.1007/s11581-019-02858-0]

[96]   Kang, W.; Pei, X.; Rusinek, C.A.; Bange, A.; Haynes, E.N.; Heineman, W.R.; Papautsky, I. Determination of lead with a copper-based electrochemical sensor. *Anal. Chem.,* **2017**, *89*(6), 3345-3352.
[http://dx.doi.org/10.1021/acs.analchem.6b03894] [PMID: 28256823]

[97]   Mahmoudian, M.R.; Basirun, W.J.; Woi, P.M.; Yousefi, R.; Alias, Y. L-Glutamine-assisted synthesis of ZnO oatmeal-like/silver composites as an electrochemical sensor for Pb^{2-} detection. *Anal. Bioanal. Chem.,* **2019**, *411*(2), 517-526.
[http://dx.doi.org/10.1007/s00216-018-1476-x] [PMID: 30498983]

[98]   Palisoc, S.; Lee, E.T.; Natividad, M.; Racines, L. Silver Nanoparticle Modified Graphene Paste Electrode for the Electrochemical Detection of Lead, Cadmium and Copper. *Int. J. Electrochem. Sci.,* **2018**, *13*(9), 8854-8866.
[http://dx.doi.org/10.20964/2018.09.03]

[99]   Baghayeri, M.; Amiri, A.; Maleki, B.; Alizadeh, Z.; Reiser, O. A simple approach for simultaneous detection of cadmium (II) and lead (II) based on glutathione coated magnetic nanoparticles as a highly selective electrochemical probe. *Sens. Actuators B Chem.,* **2018**, *273,* 1442-1450.
[http://dx.doi.org/10.1016/j.snb.2018.07.063]

[100]   Yu, L.; Zhang, P.; Dai, H.; Chen, L.; Ma, H.; Lin, M.; Shen, D. An electrochemical sensor based on Co 3 O 4 nanosheets for lead ions determination. *RSC Advances,* **2017**, *7*(63), 39611-39616.
[http://dx.doi.org/10.1039/C7RA06269A]

[101]   Promphet, N.; Rattanarat, P.; Rangkupan, R.; Chailapakul, O.; Rodthongkum, N. An electrochemical sensor based on graphene/polyaniline/polystyrene nanoporous fibers modified electrode for simultaneous determination of lead and cadmium. *Sens. Actuators B Chem.,* **2015**, *207,* 526-534.
[http://dx.doi.org/10.1016/j.snb.2014.10.126]

[102]   Yao, Y.; Wu, H.; Ping, J. Simultaneous determination of Cd(II) and Pb(II) ions in honey and milk samples using a single-walled carbon nanohorns modified screen-printed electrochemical sensor. *Food Chem.,* **2019**, *274,* 8-15.
[http://dx.doi.org/10.1016/j.foodchem.2018.08.110] [PMID: 30373012]

[103]   Bobrowski, A.; Królicka, A.; Maczuga, M.; Zarębski, J. A novel screen-printed electrode modified with lead film for adsorptive stripping voltammetric determination of cobalt and nickel. *Sens. Actuators B Chem.,* **2014**, *191,* 291-297.
[http://dx.doi.org/10.1016/j.snb.2013.10.006]

[104]   Sheikh, T.A.; Arshad, M.N.; Rahman, M.M.; Asiri, A.M.; Marwani, H.M.; Awual, M.R.; Bawazir, W.A. Trace electrochemical detection of Ni2+ ions with bidentate N, N′-(ethane-1, 2-diyl) bis (3, 4-dimethoxybenzenesulfonamide)[EDBDMBS] as a chelating agent. *Inorg. Chim. Acta,* **2017**, *464,* 157-166.
[http://dx.doi.org/10.1016/j.ica.2017.05.024]

[105]   Johnstone, R.A.W. *Mass spectrometry for chemists and bio chemists*; Cambridge University Press: Cambridge, **1996**.
[http://dx.doi.org/10.1017/CBO9781139166522]

[106]   Dass, C. *Principles and Practice of biological mass spectrometry*; Wiley: New York, **2001**.

[107]   Siuzdak, G. *Mass spectrometry for Biotechnology*; New york: aP, **1995**.

[108]   Taylor, H. *Inductively coupled plasma-mass spectrometry: practices and techniques*; academic press: California, **2001**.

[109]   D'Ilio, S.; Petrucci, F.; D'Amato, M.; Di Gregorio, M.; Senofonte, O.; Violante, N. Method validation for determination of arsenic, cadmium, chromium and lead in milk by means of dynamic reaction cell inductively coupled plasma mass spectrometry. *Anal. Chim. Acta,* **2008**, *624*(1), 59-67.
[http://dx.doi.org/10.1016/j.aca.2008.06.024] [PMID: 18706310]

[110]   Allibone, J.; Fatemian, E.; Walker, P.J. Determination of mercury in potable water by ICP-MS using gold as a stabilising agent. *J. Anal. At. Spectrom.,* **1999**, *14*(2), 235-239.
[http://dx.doi.org/10.1039/a806193i]

[111]   Liu, H.W.; Jiang, S.J.; Liu, S.H. Determination of cadmium, mercury and lead in seawater by electrothermal vaporization isotope dilution inductively coupled plasma mass spectrometry. *Spectrochim. Acta B At. Spectrosc.,* **1999**, *54*(9), 1367-1375.
[http://dx.doi.org/10.1016/S0584-8547(99)00081-6]

[112]   Suo, L.; Dong, X.; Gao, X.; Xu, J.; Huang, Z.; Ye, J.; Lu, X.; Zhao, L. Silica-coated magnetic graphene oxide nanocomposite based magnetic solid phase extraction of trace amounts of heavy metals in water samples prior to determination by inductively coupled plasma mass spectrometry. *Microchem. J.,* **2019**.104039
[http://dx.doi.org/10.1016/j.microc.2019.104039]

[113]   Su, C.K.; Chen, Y.T.; Sun, Y.C. Speciation of trace iron in environmental water using 3D-printed minicolumns coupled with inductively coupled plasma mass spectrometry. *Microchem. J.,* **2019**, *146*, 835-841.
[http://dx.doi.org/10.1016/j.microc.2019.02.015]

[114]   Chen, S.; Yan, J.; Li, J.; Lu, D. Dispersive micro-solid phase extraction using magnetic ZnFe2O4 nanotubes as adsorbent for preconcentration of Co (II), Ni (II), Mn (II) and Cd (II) followed by ICP-MS determination. *Microchem. J.,* **2019**, *147*, 232-238.
[http://dx.doi.org/10.1016/j.microc.2019.02.066]

[115]   Fang, Y.; Pan, Y.; Li, P.; Xue, M.; Pei, F.; Yang, W.; Ma, N.; Hu, Q. Simultaneous determination of arsenic and mercury species in rice by ion-pairing reversed phase chromatography with inductively coupled plasma mass spectrometry. *Food Chem.,* **2016**, *213*, 609-615.
[http://dx.doi.org/10.1016/j.foodchem.2016.07.003] [PMID: 27451225]

[116]   Zhang, D.; Yang, S.; Cheng, H.; Wang, Y.; Liu, J. Speciation of inorganic and organic species of mercury and arsenic in lotus root using high performance liquid chromatography with inductively coupled plasma mass spectrometric detection in one run. *Talanta,* **2019**, *199*, 620-627.
[http://dx.doi.org/10.1016/j.talanta.2019.03.023] [PMID: 30952306]

[117]   Chen, J.; Chen, H.; Jin, X.; Chen, H. Determination of ultra-trace amount methyl-, phenyl- and inorganic mercury in environmental and biological samples by liquid chromatography with inductively coupled plasma mass spectrometry after cloud point extraction preconcentration. *Talanta,* **2009**, *77*(4), 1381-1387.
[http://dx.doi.org/10.1016/j.talanta.2008.09.021] [PMID: 19084653]

[118]   Peng, G.; He, Q.; Zhou, G.; Li, Y.; Su, X.; Liu, M.; Fan, L. Determination of heavy metals in water samples using dual-cloud point extraction coupled with inductively coupled plasma mass spectrometry. *Anal. Methods,* **2015**, *7*(16), 6732-6739.
[http://dx.doi.org/10.1039/C5AY00801H]

[119]   Ma, S.; He, M.; Chen, B.; Deng, W.; Zheng, Q.; Hu, B. Magnetic solid phase extraction coupled with inductively coupled plasma mass spectrometry for the speciation of mercury in environmental water and human hair samples. *Talanta,* **2016**, *146*, 93-99.
[http://dx.doi.org/10.1016/j.talanta.2015.08.036] [PMID: 26695239]

[120] Zhu, S.; Chen, B.; He, M.; Huang, T.; Hu, B. Speciation of mercury in water and fish samples by HPLC-ICP-MS after magnetic solid phase extraction. *Talanta,* **2017**, *171*, 213-219.
[http://dx.doi.org/10.1016/j.talanta.2017.04.068] [PMID: 28551131]

[121] Döker, S.; Boşgelmez, İ.İ. Rapid extraction and reverse phase-liquid chromatographic separation of mercury(II) and methylmercury in fish samples with inductively coupled plasma mass spectrometric detection applying oxygen addition into plasma. *Food Chem.,* **2015**, *184*, 147-153.
[http://dx.doi.org/10.1016/j.foodchem.2015.03.067] [PMID: 25872437]

[122] Zhu, Q.Y.; Zhao, L.Y.; Sheng, D.; Chen, Y.J.; Hu, X.; Lian, H.Z.; Mao, L.; Cui, X.B. Speciation analysis of chromium by carboxylic group functionalized mesoporous silica with inductively coupled plasma mass spectrometry. *Talanta,* **2019**, *195*, 173-180.
[http://dx.doi.org/10.1016/j.talanta.2018.11.043] [PMID: 30625529]

[123] Drinčić, A.; Zuliani, T.; Ščančar, J.; Milačič, R. Determination of hexavalent Cr in river sediments by speciated isotope dilution inductively coupled plasma mass spectrometry. *Sci. Total Environ.,* **2018**, *637-638*, 1286-1294.
[http://dx.doi.org/10.1016/j.scitotenv.2018.05.112] [PMID: 29801221]

[124] Markiewicz, B.; Komorowicz, I.; Barałkiewicz, D. Accurate quantification of total chromium and its speciation form Cr(VI) in water by ICP-DRC-IDMS and HPLC/ICP-DRC-IDMS. *Talanta,* **2016**, *152*, 489-497.
[http://dx.doi.org/10.1016/j.talanta.2016.02.049] [PMID: 26992546]

[125] Mädler, S.; Todd, A.; Skip Kingston, H.M.; Pamuku, M.; Sun, F.; Tat, C.; Tooley, R.J.; Switzer, T.A.; Furdui, V.I. Ultra-trace level speciated isotope dilution measurement of Cr(VI) using ion chromatography tandem mass spectrometry in environmental waters. *Talanta,* **2016**, *156-157*, 104-111.
[http://dx.doi.org/10.1016/j.talanta.2016.04.064] [PMID: 27260441]

[126] Catalani, S.; Fostinelli, J.; Gilberti, M.E.; Apostoli, P. Application of a metal free high performance liquid chromatography with inductively coupled plasma mass spectrometry (HPLC–ICP-MS) for the determination of chromium species in drinking and tap water. *Int. J. Mass Spectrom.,* **2015**, *387*, 31-37.
[http://dx.doi.org/10.1016/j.ijms.2015.06.015]

[127] Kovács, D.; Veszely, Á.; Enesei, D.; Óvári, M.; Záray, G.; Mihucz, V.G. Feasibility of ion-exchange solid phase extraction inductively coupled plasma mass spectrometry for discrimination between inorganic As (III) and As (V) in phosphate-rich in vitro bioaccessible fractions of Ayurvedic formulations. *Spectrochim. Acta B At. Spectrosc.,* **2019**, *153*, 1-9.
[http://dx.doi.org/10.1016/j.sab.2019.01.002]

[128] Papaslioti, E.M.; Parviainen, A.; Román Alpiste, M.J.; Marchesi, C.; Garrido, C.J. Quantification of potentially toxic elements in food material by laser ablation-inductively coupled plasma-mass spectrometry (LA-ICP-MS) via pressed pellets. *Food Chem.,* **2019**, *274*, 726-732.
[http://dx.doi.org/10.1016/j.foodchem.2018.08.118] [PMID: 30373001]

[129] Zhang, N.; Shen, K.; Yang, X.; Li, Z.; Zhou, T.; Zhang, Y.; Sheng, Q.; Zheng, J. Simultaneous determination of arsenic, cadmium and lead in plant foods by ICP-MS combined with automated focused infrared ashing and cold trap. *Food Chem.,* **2018**, *264*, 462-470.
[http://dx.doi.org/10.1016/j.foodchem.2018.05.058] [PMID: 29853402]

[130] Chen, S.; Li, J.; Lu, D.; Zhang, Y. Dual extraction based on solid phase extraction and solidified floating organic drop microextraction for speciation of arsenic and its distribution in tea leaves and tea infusion by electrothermal vaporization ICP-MS. *Food Chem.,* **2016**, *211*, 741-747.
[http://dx.doi.org/10.1016/j.foodchem.2016.05.101] [PMID: 27283691]

[131] Montoro Leal, P.; Vereda Alonso, E.; López Guerrero, M.M.; Cordero, M.T.S.; Cano Pavón, J.M.; García de Torres, A. Speciation analysis of inorganic arsenic by magnetic solid phase extraction on-line with inductively coupled mass spectrometry determination. *Talanta,* **2018**, *184*, 251-259.

[http://dx.doi.org/10.1016/j.talanta.2018.03.019] [PMID: 29674040]

[132] Jia, X.; Gong, D.; Wang, J.; Huang, F.; Duan, T.; Zhang, X. Arsenic speciation in environmental waters by a new specific phosphine modified polymer microsphere preconcentration and HPLC-ICP-MS determination. *Talanta*, **2016**, *160*, 437-443.
[http://dx.doi.org/10.1016/j.talanta.2016.07.050] [PMID: 27591635]

[133] Jia, X.; Gong, D.; Xu, B.; Chi, Q.; Zhang, X. Development of a novel, fast, sensitive method for chromium speciation in wastewater based on an organic polymer as solid phase extraction material combined with HPLC-ICP-MS. *Talanta*, **2016**, *147*, 155-161.
[http://dx.doi.org/10.1016/j.talanta.2015.09.047] [PMID: 26592590]

[134] Marcinkowska, M.; Komorowicz, I.; Barałkiewicz, D. Study on multielemental speciation analysis of Cr(VI), As(III) and As(V) in water by advanced hyphenated technique HPLC/ICP-DRC-MS. Fast and reliable procedures. *Talanta*, **2015**, *144*, 233-240.
[http://dx.doi.org/10.1016/j.talanta.2015.04.087] [PMID: 26452815]

[135] Peng, H.; Zhang, N.; He, M.; Chen, B.; Hu, B. Simultaneous speciation analysis of inorganic arsenic, chromium and selenium in environmental waters by 3-(2-aminoethylamino) propyltrimethoxysilane modified multi-wall carbon nanotubes packed microcolumn solid phase extraction and ICP-MS. *Talanta*, **2015**, *131*, 266-272.
[http://dx.doi.org/10.1016/j.talanta.2014.07.054] [PMID: 25281102]

[136] He, Y.; He, M.; Nan, K.; Cao, R.; Chen, B.; Hu, B. Magnetic solid-phase extraction using sulfur-containing functional magnetic polymer for high-performance liquid chromatography-inductively coupled plasma-mass spectrometric speciation of mercury in environmental samples. *J. Chromatogr. A*, **2019**, *1595*, 19-27.
[http://dx.doi.org/10.1016/j.chroma.2019.02.050] [PMID: 30827698]

[137] Shih, T.T.; Chen, J.Y.; Luo, Y.T.; Lin, C.H.; Liu, Y.H.; Su, Y.A.; Chao, P.C.; Sun, Y.C. Development of a titanium dioxide-assisted preconcentration/on-site vapor-generation chip hyphenated with inductively coupled plasma-mass spectrometry for online determination of mercuric ions in urine samples. *Anal. Chim. Acta*, **2019**, *1063*, 82-90.
[http://dx.doi.org/10.1016/j.aca.2019.02.035] [PMID: 30967189]

CHAPTER 5

New Techniques and Methods in Explosive Analysis

Beril Anilanmert[*] and **Salih Cengiz**

İstanbul University-Cerrahpaşa, Institute of Forensic Sciences and Legal Medicine, Istanbul, Turkey

Abstract: In forensic analytical chemistry, chemical investigation of the liquid/gas/solid evidences from the crime scene after the explosion (soil, water, concrete/glass/ wood pieces, metal, clothes taken from suspects, *etc.*) and reliable identification of explosive residues on such evidences remain an active area of research due to increased demand for homeland security against terrorist and warfare threats, as well as environmental monitoring. GC-MS, LC-MS, and LC-MS/MS offer distinct advantages for laboratory analysis of explosives in post-blast samples, including soil/ water/plant matrices, *etc.* Time-of-Flight, Ion-Trap, and Orbitrap technologies provide high resolution, better analyte identification, and accurate mass information at sub-ppm levels. Direct analysis techniques, such as ambient MS has a wider range of applications and offer high sensitivity/selectivity and direct analysis from the surface of interest. Techniques like Direct Analysis In Real Time (DART) and Desorption electrospray ionization (DESI), which can ionize substances directly on surfaces, offer new opportunities for security screening of explosives. Orbitrap MS was also used together with Raman microscopy for detailed molecular-level characterization of explosives and the chemical analysis of latent fingerprints. Electro-flow focusing ionization with in-source collision-induced dissociation can be used for MS detection and chemical imaging for speciation of the signatures of explosive devices and to detect proper spatial discrimination of explosive traces. Miniaturization to be used in-field analysis with low cost is a technique work on currently. Multi-analyte detection with high selectivity/sensitivity to be able to use in as many different matrices and to be able to analyze without or with a very short and simple sample preparation methods are targeted for future analyses. Despite other reviews focussing on a certain group of techniques, this chapter summarizes some important developments in the standard MS techniques and ambient MS techniques used in laboratory, on-site, and miniaturized mass spectrometric analysis of high energetic materials in the last two decades, mostly conducted in the last decade.

[*] **Corresponding author Beril Anilanmert:** İstanbul University-Cerrahpaşa, Institute of Forensic Sciences and Legal Medicine, Istanbul, Turkey; Tel: 05387270840; Fax: 02125855387; E-mail: beril.anilanmert@istanbul.edu.tr

Atta-ur-Rahman, M. Iqbal Choudhary & Syed Ghulam Musharraf (Eds.)

Keywords: Ambient Mass Spectrometry, Confirmation, DART-MS, DESI-MS Imaging, Determination, Energetic Materials, Explosive Analysis, GC-MS, Hyphenated Techniques, Identification, Ion Trap, LCMS/MS, Mass Spectrometry (MS), Multi-Analyte Screening, On-site detection, Quantitation Methods, Recent Technologies, ToF-MS, Terrorism, Triple Quadrupole, Validation.

INTRODUCTION

In recent years, various terrorist attacks have taken place, especially in the urban cities, and unfortunately, a number of these attacks have resulted in the deaths of many innocent people [1].

Analyzing the post-blast debris for explosives in terroristic attacks helps in tracing the origin of the explosives used and the possible suspects in order to prevent further threats [1]. In forensic analytical chemistry, analysis of the liquid, gas or solid evidences (soil, water, concrete/glass/ wood pieces, metal, clothes of the suspects, *etc.*) found in the crime scene after the blast and reliable identification explosive residues on these, are very important for solving the crimes, terrorist attacks, warfares and for finding the type of the explosives and their sources. Explosive detection methods have been developed for humanitarian demining, environmental issues (since explosive residues in the environment are a threat to the human health) as groundwater and soil remediation, security screening, intelligence activities, criminal forensics, as rapid sample screening and/or quantification [2]. The results of these analyses are frequently used as evidences in courts, from the point of identification of explosives used in terrorist attacks (to identify the type of bomb), to find its country of origin or manufacturer, and to aid in connecting a suspect with the crime scene [3].

Various screening and/or quantitation methods have been developed in recent years using ion mobility spectrometry, colorimetric method, cyclic voltammetry, optical sensors, UV-Raman spectroscopy, GC-MS, LC-MS, *etc.* and new techniques as Accu-Time of Flight (TOF) Direct Analysis in Real-Time (DART)/MS and HPLC-photodiode array (PDA)-APCI negative ionization-Linear Trap Quadrupole (LTQ) MS2/Orbitrap Fourier Transform Mass Spectrometer (FTMS), also exists in literature [4 - 12]. Various sample preparation methods, such as; solid-phase extraction (SPE), solid-phase microextraction (SPME), supercritical fluid extraction (SFE), solid-liquid extraction (SLE) have been used [13 - 22]. Standard techniques for detection and quantification of explosives include gas and liquid chromatography and spectroscopy (especially mass spectroscopy) [23]. Multi-explosive identifications are achieved using these techniques. Especially in combat with terrorism, field-portable techniques gain much interest. Portable detectors are mostly based on ion mobility spectrometry and surface acoustic wave techniques. Other techniques as

Raman, terahertz spectroscopy *etc.* are also developed for remote analyses. In techniques as multiarray colorimetric detection, multiple types of sensors may be required regarding a given environment, according to a required limit of false positives, confidence needed for identification, and the molecules which are needed to be detected. Unlike the other reviews which are focused on specific MS techniques as ambient MS techniques, this chapter reviews the modern applications of MS techniques in explosive analysis from a broad scope regarding the last two decades, focusing mainly on the studies and techniques pertaining to the last decade. You can see a photo from the attack in Vodafone Arena Stadium in Istanbul in the website of European Press Agency [24].

A CONCISE INFORMATION ON EXPLOSIVES

Explosives are chemicals, having great potential energy, which is transformed into stable compounds through speedy decomposition after a sudden impact, electricity, or spark, releasing a huge sound, heat, blast, and gases [25]. Explosive materials are usually prepared from a hydrocarbon-based fuel component, and a nitrogen- or oxygen-based trigger (as nitrate or a peroxide) [26]. Explosives have

enough oxygen in their molecules to initiate and continue the very rapid progressive combustion [25]. The amount of discharged energy changes with the properties of the material, such as composition, structure, density, heat of formation and decomposition, *etc* [27]. There are two classes for military explosives as "low" or "high" explosives, depending upon the speed of propagation of the combustion reaction [28]. The rate of decomposition of low explosives is adequately slow to be used safely as a propellant in a gun. Deflagration or burning can be started with these. For high explosives, the decomposition reaction propagates faster than the speed of sound, so rapidly that can be called as "instantaneous", creating a shockwave [26, 28]. High energy materials are named as explosives, propellants and fireworks according to their properties and uses [27]. Mercury fulminate, Lead azide, octahydro-1,3,5--tetranitro-1,3,5,7-tetrazocine (HMX), Pentaerythritol tetranitrate (PETN), Amatol, Tetryl, 2,4,6-trinitrotoluene (TNT), 2,6-dinitrotoluene (2,6-DNT), Tetrytol, ethylcentralite (EC), Ethylene glycol dinitrate (EGDN), nitroglycerine (NG), 2,3- dimethyl-2,3-dinitrobutane (DMNB), Cyclonite, Tritonal, Pentolite, Ednatol, Torpex, Haleite, 1,3,5-trinitroperhydro-1,3,5-triazine (RDX), triacetone triperoxide (TATP), hexamethylene triperoxide diamine (HMTD), hexanitrostilbene (HNS), Ednatol, Pentolite, Torpex, Cyclonite, nitrate/fuel oil (AN/FO=ANFO), *etc.* are among some important high explosives [27, 28]. It is estimated there are at least 150 separate materials in use today [26].

The most common explosives used by the terrorists are high explosives such as

RDX, HMX, PETN, TNT *etc*. as shown in Fig. (**1**).

Regarding their susceptibility to detonation, explosives are investigated in three classes: primary, secondary, and tertiary. Primary explosives are extremely sensitive to friction, shock, heat, spark. Secondary explosives are usually mixed with primary explosives as well as plasticizers, waxes, or stabilizers. Organic secondary explosives are utilized in military tasks, their detonation resulting in wide spreading of poisonous residues over water bodies and the environment. Tertiary explosives don't self-explode, except in the presence of a secondary explosive. Classification of explosives, and fields in which the analysis of explosives is important, is provided in Fig. (**2**).

Highly energetic molecules, explosives quickly deteriorate *via* chemical or physical stimuli, causing a quick heat generation and gases with high pressures, like NOx, H_2O, CO, and CO_2.

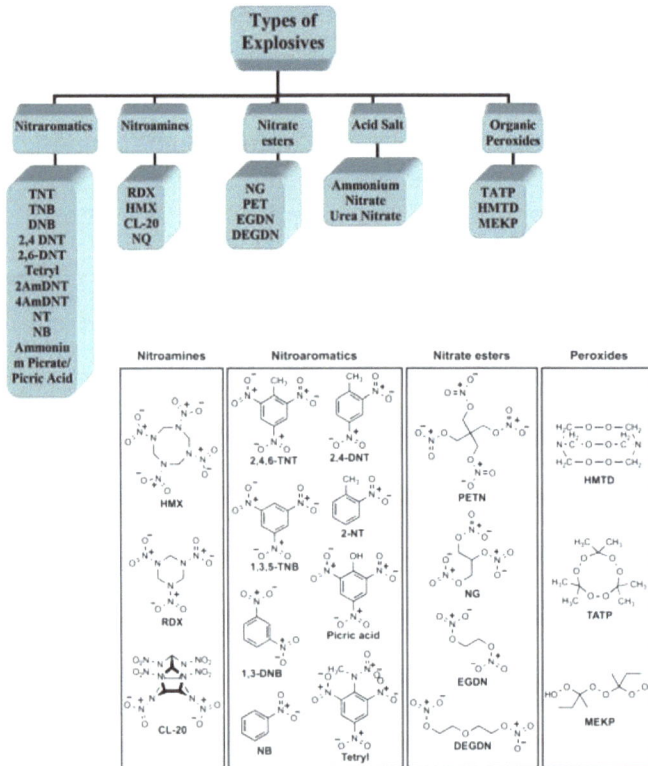

Fig. (1). The most common explosives used by the terrorists, categorized by their functional groups Modified from Ref [2] and Reprint from Ref [3].

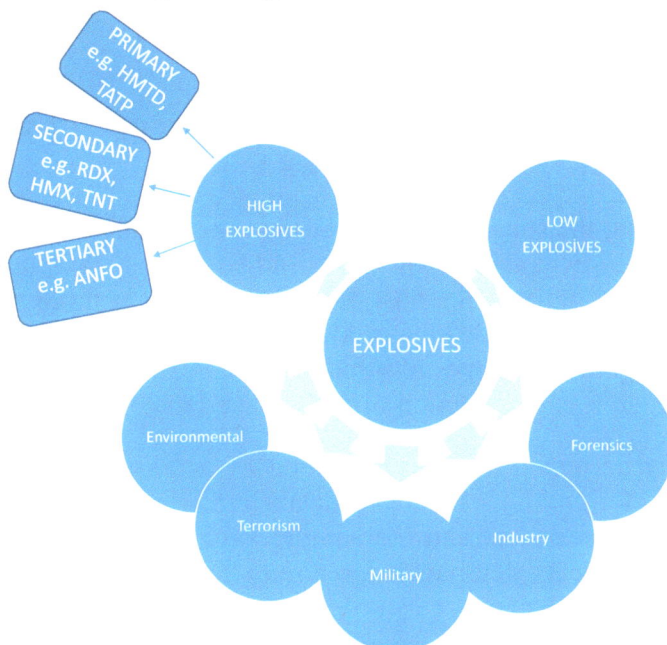

Fig. (2). Classification of explosives, and fields in which the analysis of explosives. [Reprinted from ref. 29].

Explosive systems are frequently preferred by terrorist organizations because of the possibility of an action with fewer people, having a chance to escape, and the capability of greater damage [30]. Bomb mechanisms can sometimes be detonated by using remote control devices in an LPG vehicle hidden in the roadside. In addition, the suicide bombers with their explosive devices mounted inside their clothes are one of the methods applied by the organizations. After an explosion, residues of the explosives are found, especially in the bomb assembly parts, on materials inside the explosion hole, on the surfaces close to the explosion area (vehicle explosions), and especially on thermally insulated and porous surfaces.

Improvised explosive devices (IEDs), which are criminally fabricated devices incorporating destructive, lethal, noxious, pyrotechnic or incendiary chemicals, and designed to destroy, disfigure, distract, or harass are especially tricky to detect, are encountered in numerous structures, and utilize various activation techniques [31]. They are mostly designed from commercial or homemade components. They can be employed in a broad scope of ways, normally masked as a part of the environment or as an ordinary item [26]. Explosion strategies may contain a trek wire, mobile phone, or by hand.

The explosion does not always occur on purpose [26]. Industrial explosions, for instance, are a noteworthy hazard factor in several sectors, and finding the reason for the incident is essential in preventing future disasters.

THE SAMPLE TREATMENT AND DETECTION TECHNIQUES FOR EXPLOSIVE ANALYSIS IN THE LITERATURE

After an explosion, the source of the blast and the sort of explosive utilized ought to be discovered. Because of the constrained sample amount and possible environmental contributions, the post-blast examination is more troublesome than the identification of unexploded materials in preparation [32]. Getting information about the scene, gathering materials, and using standards improve the probability of effective recognition or detection of undetonated materials.

Explosives may be transferred to the laboratory as either bulk samples or post-blast residue [31]. In the case of post-blast residues, the device may have succeeded what is intended, and the sample is collected from the scene, or the device may have been detonated for safety and its pieces with residue or swabbings may be collected.

The post-blast debris, which is insufficient size to send to the laboratory, should be packed in a gas-tight manner [30]. A swabbing with the aid of aqueous water and another swabbing with an organic solvent (methanol, acetone, *etc.*) is recommended for the evidence in a very large size, which could be a problem in transfer to the laboratory. The swab is usually cotton or gauze. The samples from evidence and the swabs taken with the organic solvent are activated with the organic solvent. The concentrated sample is obtained by evaporating the organic extract to dryness and analyzed.

Investigating a surface, and analyzing the gathered samples is generally utilized in forensic investigation, and ecological and regulatory screening [33]. Various strategies exist for gathering a targeted explosive residue, and these strategies can differ significantly, relying upon the targeted particles, the analysis technique, and the surface itself. The collection technique is centered around achieving quantitative collection and decontamination. Samples can be obtained by cutting, scraped *etc.,* when direct sampling is possible. Vacuum procedures might be utilized when contact is disadvantageous. In case of the acceptability of contact with the surface, at that point a roller, tape, or wipe can be utilized. A wide scope of swabbing materials have been utilized (cotton or rayon gauze cushions, microfiber paper, or filter paper), and in numerous techniques, a solvent is utilized in swabbing.

Identification of trace explosives, which is essential for many security screening environments, generally utilizes wipe-sampling. Current wipe-sampling assessment procedures for an effective collection of the explosive residues have some limitations: manual collection (with fingers or a wand) is constrained in its capacity to segregate a single parameter, and the slip/peel tester is restricted to a linear path. A screener who knows how to properly investigate and choose the right surface for sample collection will be much more successful at detecting trace explosives.

A recently created wipe-testing instrument, using a commercial off-the-shelf (COTS) 3D printer repurposed for its XYZ stage permitted, automated for two-dimensional wipe-examining patterns to be investigated, with consistent power and speed of accumulation through the length of the sampling path [33]. This new technique isn't just equipped for researching parameters of the current technology (wipe materials, test surfaces, powers of accumulation, and straight example designs), it has also added capacities to examine extra parameters, as directional wipe paths (for example "L" and "U" shapes, square, and serpentine), allowing numerous lines to be inspected during a single collection without the requirement of any change in the instrument. Parametric examinations were performed utilizing 1,3,5-trinitroperhydro-1,3,5-triazine (RDX) for various scenarios. ABS plastic and ballistics nylon texture (to imitate luggage), cardboard and wrapping tape (to imitate cargo), synthetic leather (to simulate luggage and dresses), and stainless steel (to mimic vehicles) were used as matrices. RDX solution was imprinted onto PTFE thin films utilizing drop-on-demand (DOD) inkjet printing. After drying, RDX residues were transferred to a test surface reproduced with genuine deposits found in field wipe-sampling situations. Through a wipe-sampling, the RDX residues were collected onto a wipe. Then, RDX was extracted from both the thin film and the wipe and analyzed using electrospray ionization-mass spectrometry (ESI-MS) to find out the success of transfer and collection efficiency (CE%), in order. The placement of the wipe mount eliminated a possible user error and ensured a smooth translational movement while keeping the speed, power, distance and direction constant. These improved the reproducibility of the CE% over the currently utilized TL-slip/strip analyzer while keeping the variations under 4% . It could be used for different types of wipe materials, test surfaces, and loads. The additional xy-stage features provided a wider variety of wipe sampling patterns (straight, serpentine, and square) just as providing multiple lines without the requirement for user modifications.

High explosives as nitroaromatics, nitrate esters, nitramines, or mixtures of these with or without other ingredients cause prosperity and health risk to individuals and the earth, even at low parts per billion measurements [34, 35]. Thus, the method development studies reporting environmental aspects of explosives are mainly focused on these. For example, TNT is mutagenic, causes liver damage and anemia, and prevents the growth of certain fungi, yeast, and bacteria [35].

Trace levels of explosives identification are very important in areas, which are under suspect of contamination, for finding out the extent of the contamination before remediation, observing groundwater quality, and preventing toxicity to humans, animals and plants.

Explosive screening studies are conducted in two categories, one of them is fast identification (for tactical, field, and evidence screening applications) of explosives, and the other is sensitivity and quantification studies (for probative criminal forensic) [36].

For fast, automatic, and non-contact identification of trace amounts of explosives for homeland security and environmental safety, various spectroscopic technologies have been developed; for example, terahertz (THz) spectroscopy [37], laser-induced breakdown spectroscopy (LIBS), infrared (IR) and Raman spectroscopy [29, 38 - 40], ion mobility spectrometry (IMS) [41], nuclear magnetic resonance (NMR), nuclear quadrupole resonance (NQR), laser-induced thermal emissions (LITE), mass spectrometry, optical emission spectroscopy (OES), photo-thermal infrared microwave, and millimeter-wave, *etc* [41 - 43]. Electromagnetic radiations as X-rays, γ rays, and UV are also utilized in explosive detection. Significant improvements have been achieved in such spectroscopic techniques (THz spectroscopy, LIBS, Raman spectroscopy, and IMS) in explosive detection. Array-based sensing techniques are also developed to discriminate single analyte among multi-analytes and allow to detect the complicated mixture of analytes [44]. However, the utilization of sensor arrays for the detection and identification of various explosives is very limited. Up to now, sensor arrays designed for explosives have been manufactured using fluorescent polymers, semiconductor quantum dots, tricarbazole-based nanofibers, and nanoribbons and silver-based surface-enhanced Raman scattering. Their problems are the need of an expensive instrument, tiring sample preparation, and laborious sensor array manufacturing, unsatisfactory reproducibility/accuracy. Their advantages are high sensitivity, simple operation, low cost, and fast analysis. A multichannel colorimetric sensor array exists in the literature using indigo carmine reduction with DTT. It achieved to detect 10 explosives, including PA, DNT, NTO, CL-20, LLM-105, RDX, FOX-7, NG, NP, and DNP.

Recently, considerable progress has also been made in the field of explosive detection based on the fluorescent sensors of quantum dots (QDs) with a high surface-area-to- volume ratio [45]. In comparison with organic dyes, such as rhodamine, the fluorescent quantum dots are 20 times as bright and 100 times as stable against photobleaching, showing better potential applications in various fields. Sensitivities of the QDs sensors are controlled by the electron-donating properties of the capping layer that modifies the particles, thus allowing the quantitative analysis of the explosive substrates. A multichannel fluorescent sensor array based on nanofibrous membranes loaded with ZnS QDs featuring several surface ligands, providing a synergistic effect with the high surface area-to-volume ratio of QDs, the good permeability of nanofiber membranes and the differential quenching introduced by surface ligands was created and demonstrated for the discriminative detection of dinitro-toluene, TNT, picric acid, nitrobenzene. The vapors of these nitroaromatic explosives could reliably be detected and discriminated by the array at room temperature. The sensing system is designed to respond to a range of explosives through supramolecular interactions, such as host-guest binding and electrostatics, causing fluorescence quenching of the QDs, to create an analytical fingerprint for the sensitive, quick, discriminative detection of explosive vapors at room temperature. It could, for example, be coupled with the technology of image recognition and large data analysis for a rapid diagnostic test of explosives.

After an explosion, the primary objective of law enforcement organizations is to identify the explosives. Chemical profiling to differentiate between samples or tracing the explosive to its source (for example, to define a single batch, a single manufacturer, or a geographic region), are carried out additionally in forensic laboratories. Generally, the secondary analyses are most effective in the identification of unexploded molecules, as in stable isotope measurement technique. Isotope ratio mass spectrometry (IRMS) depends on the natural distribution of common stable isotopes in various places on the earth or from various sources. Isotope ratios can present an elemental signature for the sources of explosives [32]. Benson *et al.* [46], explored the nitrogen isotope variability within a time series of ammonium nitrate (AN) production at three factories. After that study, Grimm *et al.* [47] used the variation in the oxygen isotope composition of AN, as well as the nitrogen isotope composition of NO_3^- and NH_4^+ for the first time to differentiate between unexploded AN derivatives. The use of stable nitrogen and oxygen isotope ratios ($^{15}N/^{14}N$ and $^{18}O/^{16}O$, expressed as $\delta^{15}N$ and $\delta^{18}O$) of AN in forensic investigations can provide supplemental properties for comparison in materials which are chemically indistinguishable with other analysis methods. The importance of this method depends on the variety between various sources and the insignificant variations within a source. Furthermore, since AN has a modest isotopic range, a potassium nitrate precipitation technique was developed to separate the ions (NO_3^- and NH_4^+) for individual δ15N analysis and increased the identification power. So that the developed method isolates and isotopically analyzes the NO_3^- ion *via* potassium nitrate precipitation, besides the standard nitrogen and oxygen isotope analyses of a bulk AN sample.

For a healthy, stable isotope characterization to be useful in sample attribution and differentiation, a proven variety between different sources and minimal variation within a source should exist. A small range of isotopic ratios are produced in the manufacture of AN, and it makes the interpretation hard. The variability in isotopic ratios could be beneficial in the comparison of AN among two or more bombs or a bomb, and a concealed AN caught on a suspect. However, lot-to-lot differences of one producer could confuse the results. In a developed method, samples were dissolved in 18.2 MΩ-cm water and filtered through a 0.45 μm nylon membrane. The samples were then dried under N_2 and stored in a desiccator. Then, absorption filtering was performed. Since nitrogen resides in two positions in AN, a method with KOH was employed to separate the two-component ions prior to isotope ratio analysis (To isolate the NO_3^- ion by precipitation as potassium nitrate (KNO_3)).

X-ray diffraction (XRD) is a less utilized kind of gadget for explosion investigation. XRD of a sample gives crystalline structure data. It is a promising non-intrusive, non-contact technique to recognize and distinguish fluid and strong materials X-ray diffraction, which provides crystalline structure information on a sample, is a less utilized technique for explosive analysis. It is a non-invasive, non-contact technique to identify liquids and solids [48]. In energy dispersive X-Ray Diffraction Spectrometry (ED-XRD) a polychromatic X-ray beam is used to observe the interactions between photon and substance during scattering. An energy-resolved detector at a fixed scattering angle measures the material-specific spectrum. The specificity of the spectrum is due to the atomic planar spacing (d) and radiation wavelength (λ) of the material, according to Bragg's law ($2d\sin\theta = n\lambda$, n is an integer). Also, as in angle dispersive X-ray diffraction (AD-XRD) λ can be fixed, and θ is varied during analysis. The equipment is not portable because of its size and shielding requirements. Basic component and discrimination analysis differentiate between inert materials and explosives, *e.g.* Semtex and ammonium nitrate emulsions.

Neutron activation and gamma emission active neutron interrogation methods are also utilized to find out the relative chemical content of certain elements (N, O, Cl, H) through their characteristic gamma-ray emission, neutron scattering, and neutron absorption. Typically, the high nitrogen content is used to detect an explosive. Neutron activation works well for high nitrogen explosives. In neutron activation analysis, neutrons can pass through high atomic number materials, such as metals, and there is a specificity to differentiate between organic and inorganic materials. However, it has also disadvantages as producing harmful radiation doses if it is not shielded properly or at the correct stand-off and neutron interrogation can cause unwanted material activation. Developments in this area are mostly focused on the hyphenating of different techniques, advances in neutron sources from the point of energy and portability, and increase of detection power. In a recent study, thermal neutron analysis (TNA), fast neutron analysis

(FNA), and dual x-ray imaging were hyphenated to analyze ratios of Cl, H, Fe, N, where the use of different ratios of the elements (*e.g.*, N/H, Cl/H, Cl/N) reduced the false alarm rate [49].

Terahertz (THz) technology, which detects explosives of various sources, has attracted much interest [2]. THz spectroscopy can penetrate and detect non-metallic items, hidden or concealed objects [48]. There is a serious effort on obtaining spectral signatures of explosives, as well as developing signal processing techniques for both spectroscopy and imaging. There are few THz detection studies containing quantitative analytical parameters. The selectivity of the THz spectral signatures and their variation with sample type or matrix are evaluated in many papers. A common technique for THz spectroscopy is Time-Domain Spectroscopy (THz TDS), a pump-probe method. A THz pump pulse (produced by ultrafast laser) with a shorter probe is sent through a sample to a detector. The THz electrical field can be recorded in a time period and can be Fourier transformed into a frequency spectrum. Many studies have been performed on the spectral signatures of explosives and related materials using THz TDS. The spectral signatures of explosive mixtures and plastic compositions, homemade explosives, and related materials have also been investigated. Even stand-off detection using THz TDS has been studied. THz spectroscopy presents opportunities for the detection of explosives, concealed in e. g. envelopes, clothing, *etc.*, or while in interference with a barrier, or in the presence of surface roughness of the samples, and water vapor/humidity in the air. These challenges were mentioned in the literature, and new methods are developed to separate spectral fingerprints of explosives from barrier materials, and signal processing algorithms were produced to compensate for water vapor absorption. However, these new methods have been only partially successful because current THz spectroscopy techniques allow the collection of spectra in a narrow THz window (3–6 THz). This range is not adequate to discriminate different explosives and their matrix separately, since many compounds may appear similar. A larger THz bandwidth is expected to identify the sole features in the spectrum specific to each explosive. Reliable measurements with a larger scale THz bandwidth requires a more sensitive arrangement and careful signal processing, since the reflection of hazardous materials is weak, and the effect of surface quality is high. To prevent peak broadening, a waveguide-based TDS technique can be used where the crystal plane is highly directed to the THz radiation.

Ion mobility spectrometry (IMS) is a widely used technique in the detection of explosives before the blast [50]. IMS has advantages like firmness, portability, and on-site use. However, it has also disadvantages as low resolving power, limited selectivity, and chemical interference, so they are not always directly applicable to the comprehensive screening of various explosives. For the ion mobility spectrometry, future development directions are miniaturization, the substitution of non-radioactive ionization sources, and improvement of instrument performance.

Hyphenating various methods by sharing lasers, spectrometers, and optical paths, different spectral methods, is an effective strategy to provide accurate identification. (Forexample Raman and Lazer Induced Breakdown Spectroscopy (LIBS) can be combined in a single instrument, and trials are carried out to reduce the size and complexity of the instrument. LIBS yet has some disadvantages to overcome. Interference of air oxygen and nitrogen is a problem, and a system designed to prioritize remnants can help eliminate peripheral interference. Furthermore, the dual pulse LIBS can decrease the effect of the ambient atmosphere.

New analytical methods are required for faster and more sensitive detection. Rapid real-time analysis of the trace amount of explosives with high precision, resolution without the requirement of long procedures for sample preparation, new and reliable methods in the laboratory applications are targeted. At present, the most advanced techniques are spectroscopy-based detection techniques, and most probably, it will be so also in the future. The most important principles are improving the sensitivity and specificity /selectivity for these molecules.

Because of its nondestructive nature, among these, Raman is one of the advantageous techniques in explosive analysis in recent years. Raman spectroscopy investigated vibrational-frequency modes, thus, the Raman signal of the explosive gives a characteristic spectrum or "molecular fingerprint" [51]. When combined with nanosensors, such selective technologies have huge power in analytical chemistry, especially in homeland security applications, to fight terror and military threats. It is possible through combining SERS nanosensors with Raman Spectrometry, it amplifies the weak Raman signal of a molecule by contacting it with metallic nanostructures, as well as eliminating the strong background fluorescence. At 2018, Lianage *et al.* [51] have developed a Surface Enhanced Raman Spectroscopy (SERS) nanosensor using self-assembling chemically synthesized gold triangular nanoprisms (Au TNPs), which provide strong electromagnetic field enhancements at the sharp tips and edges on a pressure-sensitive flexible adhesive film (which is sold in the market). Explosive analytes may be dropped as a solution or transferred from a thumb impression directly on a nanosensor. SERS spectra are collected through a Raman spectrometer using diode laser excitation adjusted to 785 nm. SERS substrate here is the nanostructure, and the localized surface plasmon resonance (LSPR) properties of the nanostructure, which arise from the resonance frequency of the total oscillations of its conducting electrons with that of the novel SERS nanosensor, amplifies the SERS signal. Low LOD values up to ppq levels may be achieved for common military high explosives with this technique. The features of LSPR produce strong electromagnetic field enhancements ("a large number of hot spots") around the nanostructures, and electrical fields with higher amplitudes seem to be the basic contributor to SERS sensitivity enhancement. Many flexible (*e.g.* cotton or paper-based) nanostructure SERS sensors have been manufactured for explosive analysis with good sensitivity, but the lack of appropriate immobilization of 3D nanostructures onto the surface of SERS sensors prevents the efficient sampling from jagged or uneven surfaces. Moreover, the swab-based methods which require swabbing and solvent-extraction could cause changes in sample characteristics, resulting in a change of fingerprint impression. Current portable Raman spectrometers usually do not combine the microscope, so they can only be applied to the analysis of unexploded explosives in high amounts. Further research through the use of portable Raman microscopes will also be useful in smaller amounts.

Also, colorimetry-based commercial off-the-shelf (COTS) explosives detection devices are used in the explosive analysis. These are based on methods as adding reagent drops to sample swipes, which are previously impregnated with chemicals, or mixing the sample and these chemicals in a solution [50]. Current COTS devices provide sensitivity for a broad range of explosives at <mg levels, and identification by class (organic nitrate or nitramine, inorganic nitrates, chlorates, *etc.*) is possible. They are inexpensive and easy to perform with and useful in field testing. Multiple colorimetric tests on a single test paper with isolated reagents speeds up instant field testing. There are also recent techniques as colorimetric sensor arrays, smartphone-based camera analysis, and synthesis of new test reagents, especially with nanoparticles. In colorimetric sensors for multiple explosives, the color changes of several chemically different sensors are affected simultaneously, the pattern of color changes instead of a single specific color change is used for detection [50]. This provides high selectivity for a broad range of detected molecules, with properly selected multi- sensors. The use of such sensors has been shown for vapor-phase detection of volatile organic chemicals, toxic industrial chemicals, commercial products, *etc.* Some of these methods are shown to achieve even low ppb values [51 - 54]. Current forensic methods used in producing results for the courts require some standard instruments as GC-MS, GC-MS/MS, GC-NCI, GC-ECD, LC-MS, LC- MS/MS, and ion chromatography (IC), *etc* [55, 56]. to cover as much as possible explosive molecules, and to keep analysis times averagely several minutes per sample [49]. TLC, HPLC [57], GC-TEA, GC-NPD [34] are also among the methods that exist in the literature. Techniques as Raman and IMS are utilized in remote detection techniques, but MS techniques have more advantages as high accuracy, high sensitivity, selectivity/specificity, and less false negatives. Chromatography is one

of the most preferred techniques because of the chance to detect, identify, and quantify the explosives molecules separate from each other and the interferents from the matrix. HPLC-UV is utilized in the analysis of organic explosives, sensitivity can decrease to ng levels, but for the sake of selectivity, chromatographic techniques with MS detectors take its place in recent years. GC-TEA is a very selective but expensive method used in the analysis of nitro compounds and may be sensitive to pg levels [30, 34]. The explosives analyzed with GC methods should be volatile and should not be heat-labile. The NPD is less selective than the TEA, but insensitive to nitrate esters. GC-ECD is less selective and is more sensitive for nitroaromatics (in pg levels) than the TEA or NPD [30, 34].

Liquid chromatography provides the opportunity to analyze directly and simultaneously a wide scope of analytes, including the thermally unstable compounds without the risk of degradation, which is possible during GC analysis [58].

Other separation devices, such as capillary electrophoresis (CE), also show promising results for explosive analyses [59]. CE has a simple instrumental setup and great versatility, it is economical and requires solvent and sample in lower amounts. It has high on-field analysis potential to avoid transportation of the sample. Various kinds of detectors can be used in combination with CE as indirect photometric detection [60, 61], laser-induced fluorescence [62 - 64], MS [65], and C4D [1].

Various sample treatment techniques are used in the literature for explosive analysis [66]. Post-blast residues of organic explosive compounds are difficult to detect because a detonation causes the traces of explosives spread over a large area, where many interfering compounds and a complex matrix is formed after the explosion. Sample preparation techniques such as solid-phase extraction (SPE) [67], liquid-liquid extraction (LLE) [68] solid-liquid extraction (SLE) [69 - 71], solid-phase microextraction (SPME) [20, 72, 73], and single drop microextraction (SDME) [74] are used to improve the isolation of explosive compounds away from the debris matrix.

EPA Method 8330B uses HPLC and requires 2.0 g soil and 10.0 mL of acetonitrile for 18 h extraction [75].

Water analysis requires more accessible procedures than samples as soil, concrete, *etc*. The water matrix may even do not require a pre-concentration step, as in the study of Schramm *et al.*, who has developed an LC/MS method for DNB, TNB, TNT, DNT, HMX, RDX and PETN analysis in groundwater samples from a military site. Samples were filtrated through syringe filters and placed directly into the autosampler [58]. Mu *et al.* (2012) followed the same approach of filtration using a 0.22 μm nylon membrane, with direct injection into the LC/MS system [76]. LLE was also used for TNT and RDX in the literature [77]. Other sample preparation steps include centrifugation, solvent exchange, evaporation, distillation, desalting, microdialysis, freeze-drying, sieving, and grinding (for solid samples) [78].

Conventional SPE, LLE, supercritical fluid extraction (SFE), or supercritical fluid chromatography (SFC) and column chromatography are among the frequently used extraction/separation methods for liquid samples as well as various matrices. Ultrasonic assisted extraction (UAE) technique is frequently employed for soil and sediment extraction [69 - 71]. In 2011, an ultrasonic titanium probe was used through immersing into a reaction mixture of water sample and extracting solvent for the extraction of NB and TNT [79].

In 2013, Detata *et al.* [80], carried out some studies on spiked solvent, using various SPE cartridges and adsorbents for column clean-up. Silica-based cleanup cartridges did not require sample pretreatment and the process lasted less than 10 min. Average recoveries were around 87% with RSD% values less than 4%. Oasis HLB fabricated of polyvinyl-pyrrolidone-divinylbenzene (PVP-DVB) copolymer by Waters Corporation displayed better results when compared with Porapak RDX and silica gel cartridge. Clean-up with Alumina/Florisil column has shown recoveries greater than 75% for all explosives analyzed, except TNB and HMX.

The recovery obtained with Oasis HLB was between 78% and 124% for DNB, DNT, TNT, HMX, RDX and PETN. These recoveries belong to the post-elution step, without any evaporation. NB seems to be lost after evaporation (because of volatilization) and can not be determined at HPLC/UV. The effect of solvent volumes were also studied and explained in detail in the study. A variety of sorbents, such as silica based, graphitic carbon, copolymers and alumina, were tested for explosive analysis and recoveries reported in a study in 2017 [81], SPME extraction technique has been later evolved to planar solid phase microextraction (PSPME) through maximising the surface area of the SPME fibres of a planar geometry [82, 83].

A clean-up method in soil using SPE has been also developed for 12 nitro-organic explosives [84]. A large scope of explosives or explosive-related compounds were evaluated, including nitramines, nitrate esters, nitroaromatics, and a nitroalkane. Spiked soil samples were initially extracted with acetone. Then the acetone extract was separated from the soil using a pipette and diluted to 50 mL with water. SPE was applied to this solution, and then GC-ECD was used for analysis. EmporeTM SDB-XC, Oasis1 HLB, and Bond Elut NEXUS cartridges were compared as SPE sorbents. The NEXUS cartridge provided the best recovery for 12 explosives in soil (average 48%) with the shortest processing time (<30 min). At the end of the validation; 12 compounds were detectable at 0.02 mg/g or lower in soil, sand, and loam over 3 days. In the most complex matrix, which is processed loam; 7 explosives were stable up to 7 days at 2 mg/g and 3 were stable at 0.2 mg/g. When authors compared the SPE literature methods; the highest recoveries belonged to the Oasis HLB cartridges, but the extraction time was long because of large volumes used in conditioning, low flows, and clogging. The procedure with Empore SDB-XC cartridges needed a 1 - 2 h time, but no improvements could be achieved in recoveries and times even higher vacuum was used . The high cost of the SDB-XC cartridges is also a disadvantage. The process can be completed with Bond Elut NEXUS cartridges within less than an hour and good recoveries can be obtained using an optimized extraction method.

Ionic and polymeric ionic liquids (ILs and PILs) are widely utilized in various applications. They also serve as dispersants in the extraction of DNT and TNT, through SPME, from water samples. An imide-based IL is used as an extracting solvent, producing very good recoveries as 97–101% for TNT and DNT in samples of surface water [85]. The recoveries are related to NaCl concentration, the mass of IL, the type, and the volume of the dispersant. LPME, SPME and cloud-point extraction (CPE) techniques using surfactants are also used for explosive extraction in the literature [78]. Babaee *et al.* [86] have utilized CPE to extract TNT, HMX, RDX and PETN from the river and well water samples in Iran. This new method displayed better recoveries (range 97–101%) than EPA methods, lower LODs than SPME HPLC/UV (0.08–0.40 µg/L) and comparable LODs to SPE HPLC/UV (0.09–0.29 µg/L). Electrospun nanofibres also have attracted attention recently. Fluorescent films are manufactured in the laboratory by electrospinning pyrene-doped polymers dissolved in DMF/THF, which are targeted for sensitive and selective determination of explosives in aqueous solutions. DNT, TNT and RDX could be detected in nanomolar levels [87].

The recent trend in sample preparation techniques is solvent-less applications, for example LLE has been scaled down to LPME (liquid-phase microextraction) with the use of microvolumes of extraction solvent, and SPME has been developed as another solvent-less extraction method for trace concentrations of explosives [78]. LLEs include microtechniques as membrane-assisted, single drop, fibers, ionic liquids, drop in the drop, continuous flow, vortex/microwave assisted, directly suspended and floating drop, where the ratio of sample/solvent is very high. Enrichment factors can reach numbers like 100. Membrane-assisted, salting-out, electrochemically modulated, and cloud-point extractions are also among the solvent-less techniques. Similarly, under the SPE umbrella, techniques as stir-bar, thin-film, in tip, fibers, in tube syringe may be utilized, along with various sorbent types and other techniques as matrix dispersion and Quick Easy Cheap Effective Rugged Safe (QuEChERS) method. SPME can be automated and combined with LC/MS or GC/MS [78]. The fibre may be immersed into a liquid in order to adsorb the analyte molecules or placed in the headspace of the sample vial with a sample to adsorb the volatile analytes. Volatile analytes can be thermally desorbed inside a GC injector or given into a mobile phase of an LC system [78]. LPME and SPME are widely utilized for the extraction of NB, DNB, DNT, TNB, and TNT from soil and water samples. Types of these techniques are explained in a review of Padron *et al.* [88].

SPME is a good alternative to conventional techniques in the analysis of volatile or semivolatile organic compounds [89]. It is fast, simple, precise, sensitive, and no solvent is needed. In a study performed by Psillakis *et al.*, SPME technique was used for extraction from soil matrices where the disadvantages of solvent evaporation are eliminated. A polymer-coated fiber is utilized to extract volatile and semi-volatile analytes from a solution. The quantity of analyte extracted is determined by a partitioning equilibrium established between the polymer coating on the fiber and the substrate being sampled. In the method developed; 2.00 g of sample was placed in 10 mL vials and 8 mL 27% (w/v) NaCl solution in water was added. The spiked samples were sonicated for 2 h. The soil-slurry sample was kept in the dark overnight. 5 mL of supernatant was removed, placed in 10 mL vials and internal standard was added SPME was performed with a 65 μm polydimethylsiloxane /divinylbenzene (PDMS/DVB) SPME fiber. The solution was stirred rapidly and consistently with a magnetic stir bar, and then the desorption from SPME fiber was carried out at the GC-MS injector using splitless mode for 5 minutes just before GC-MS analysis with positive electron impact mode. Ion source was kept at 200 °C and the solvent delay at 4 min. The injection port temperature was 250 °C. 30 m x 0.25 mm x 0.25 μm SPB-1701 capillary column was used for separation. The head-pressure was 60 kPa. The column oven was initially at 100 °C for 2 min, then increased to 200°C with a rate of 10 °C min⁻¹, then to 250 °C with 20°C min⁻¹ rate and held for 5 min. Helium was used as carrier gas (0.9 mL min⁻¹, at 100°C). The authors have found out that although the headspace shows good performance for explosives with lower molecular weight (NB and NT derivatives), extraction of the larger molecules (DNB and DNT derivatives) was not complete, or even not possible (ADNT derivatives, TNT and TNB) even in prolonged times. As a result, the sample preparation method produced low recovery for all targeted explosives (2-NT, 3-NT, 4-NT, 2,6-DNT, 1,3-DNB, 2,4-DNT, 2-ADNT, 4-ADNT) probably because the water was used for extraction.

New fibre coatings are developed to increase sensitivity in the extraction of explosives, along with more powerful methods of analysis. Sol-gel technology and carbon nanotubes (CNTs) offer interesting perspectives for DNB, TNB, TNT, and TNB identification and quantitation [90]. Hybrid materials of polymers (active phase) and magnetic nanoparticles (solid support) are developed to increase the recovery of the sorbent and to make the sample preparation easier. Dispersive micro-solid phase extraction (D-μSPE), which combines polymeric microparticles and magnetic nanoparticles is used successfully in surface water samples for nitroaromatics (NB, DNB, DNT) [91]. The most recent improvements in D-μSPE, manufactured from sorbents made of nanomaterials, magnetic nanocomposites, hybrid nanomaterials and a comparison to SPE is presented in a review paper [92]. This technique could also be used in post-blast explosion water samples.

New, efficient, and inexpensive analytical methods are required for environmental monitoring and evaluation [93]. So it's seen from the literature that focus from adsorbents such as SPE is currently shifting to sorption materials [78]. Combining sample extraction, purification, and enrichment, utilizing sample preparation techniques as solid-phase microextraction (SPME) and stir bar sorptive extraction (SBSE), magnetic solid-phase extraction (MSPE) carries sample preparation to a more green approach [78, 93]. Various materials as metal-organic frameworks, nano and micromaterials are started to be tried in sorptive extraction techniques, as another advantage over other extraction techniques [94]. SPME and SBSE are commercial micro-extraction extraction techniques based on analyte adsorption as well as gum-phase extraction and equilibrium gum-phase extraction [95, 96]. Sorptive sampling techniques combined with GC-thermal desorption (GC-TD), are more often being used to overcome sensitivity limitations of liquid extraction techniques because of injecting only an aliquot of the extract with low solvent into the analytical instrument [95, 98]. Sample handling and solvent consumption are reduced while the concentration factor is equal or higher than with classical liquid–liquid extraction. Since they present satisfactory reliability and robustness for routine sample preparation, sorptive extraction techniques started to be accepted into official methods. In TD-GC methods, the stir-bar is inserted in the heated GC injection *via* TD and transfer the desorbed analytes to the GC column [97]. In TD-GC, both volatile and semi-volatile molecules are desorbed from the sorbent, and are transferred directly to the hot GC injection port [99], or in a stainless-steel or a glass tube [100, 101]. Heat is applied after the introduction of the analytes to the GC injection port through a heated transfer line of TD. This solvent-free method may be automated [102]. Liquid desorption (LD) is the alternative when thermally labile solutes are analyzed [97] and the separation is carried out using liquid chromatography (LC). In liquid desorption, the stir bar is placed in a small amount of a proper solvent (GC) or the mobile phase (LC). LD methodologies have high sensitivity and reproducibility. However when SBSE is used along with LD technique, it has limitations related to the manual performing of stir-bar removing from the sample, rinsing and drying. The main drawback of the SBSE technique is the desorption step, especially to LC, because of the complexity in the automation 2D GC (GC-GC) is a strong technique in the detection of a wider range of trace analytes in a complex matrix during a single run [103]. Better resolution (increased selectivity), higher sensitivity is achieved compared to the traditional GC [104, 105]. Wooding *et al.* used GC-GC-TOF/MS in the separation and detection of explosives in a study and developed a low-cost, disposable/reusable, in-house built sorptive extraction sampler alternative to SBSE using a twister sorptive sampler sold in the market, where polydimethylsiloxane (PDMS) polymer was used in both as sorption [93]. The performance of TD with disposable/reusable samplers in the inlet of a GC was compared to that of conventional thermal desorption system (TDS), which is sold in the market. Ten micropollutants, from a range of heterogeneous compounds, were selected as a model. The low-cost and disposable in-house manufactured sampler gave results comparable to commercial SBSE. Direct thermal desorption of the disposable sampler in the inlet of a GC eliminated the need for expensive consumable cryogenics, and analysis time was greatly reduced as a long desorption temperature programming was not necessary. The results of in-house developed PDMS loop and SBSE did not differ significantly. Therefore, the PDMS loop is suggested to be a cost-effective (with less than 1 $) alternative to SBSE. The disposable extraction apparatus is claimed to eliminate the potential analyte carry-over.

PDMS loop sampler, is suggested to give results that do not differ significantly from a loop sampler desorbed in a commercial TDS when desorbed directly in an inlet of a GC. Thermal desorption directly in the GC inlet, reduced sample introduction time, and cryo-focusing was not necessary. The easier sample introduction is related to the low thermal mass of PDMS loop sampler, which facilitates rapid desorption and provides narrow peaks for volatile and semi-volatile compounds even with the splitless mode. Such inexpensive and in-laborious systems may be evaluated by researchers for future explosive analysis development studies with GC-MS techniques.

MASS SPECTROMETRIC TECHNIQUES USED IN EXPLOSIVE ANALYSIS

Mass spectrometry (MS) has become the most successful technology for the analysis of such compounds in water, soil, *etc.* throughout the world. Recent improvements in MS analysis help to understand the ion behavior of explosive molecules and the influence of additives and reagent gases on the formation of ions [35]. Original MS approaches have been developed especially for fast analysis of explosives. Various types of MS techniques are successfully used in the detection of trace levels of explosives in post-blast samples, including soil, water and plant matrices, *etc* [106]. There are also current quantitative methods approved by the Environmental Protection Agency (EPA) for explosives as in organonitrate explosives detection with Methods 8095 and 529 which uses GC with electron capture (ECD) and MS detection, respectively, and Method 8330 performed with HPLC-UV [36]. LOQ values for these methods change from 0.084 ng/mL to 13 ng/mL for the analysis times ranging from 1.2 min to 14 min. GC/MS is one of the most preferred techniques for the determination of certain organic explosives [34]. Various determination methods for explosive molecules have been developed for humanitarian demining, groundwater and soil remediation, security screening, tactical forensic and intelligence, criminal forensic, rapid sample screening, and trace-evidence chemical imaging [36].

Analysis with GC-MS & GC-MS/MS

In GC techniques combined with MS, GC separates the volatile and thermally stable analytes in a sample, and MS fragments the analyte for identification of the analyte regarding the fragments according to their mass/ electrostatic charge (m/z) ratios [107]. The elution time in GC is not sufficient to identify the detected compounds where the identification depends on retention time matching, which may be inaccurate or misleading. The molecular ion or the fragment chosen in MS detector is enhanced through the filtration of the other fragments, which may cause background noise, and this improved confidence in sample identification and improved sensitivity, particularly for compounds that are hard to analyze. GC-MS also significantly increased the range of thermally labile and low volatility samples amenable for analysis, besides many options that provide reasons to use the GC-MS in a broad range of areas.

Up to now, ionization techniques as EI, PCI, and NCI were used in GC/MS analysis of explosives in literature and case-related samples [108]. In the early 80's, negative chemical ionization (NCI) MS mode in GC-MS, was proposed as a more sensitive and informative method for most explosive identification, than the positive chemical ionization (PCI) and the common electron impact (EI) modes [109]. Currently, electron impact (EI) and chemical ionization (CI) techniques are generally used for ionization and fragmentation in GC-MS [107]. In the literature, thermal desorption GC-NCI/MS coupled with ECD (Electron Capture Detector) was also shown to analyze trace explosives in post-blast debris [109]. Then, tandem MS was used for improved analytical selectivity and demonstrated for nine explosive compounds in interfering matrices. GC-MS methods are used for the identification of explosives from different environmental matrices, such as water, soil and air in time; however, they generally still lack the desired level of trace sensitivity. GC-MS/MS contributes as superior performance to selectivity in GC-MS methods, since it can differentiate the analytes from interferences and co-eluting compounds, and increase sensitivity since the background noise in a post-explosion sample is reduced during analysis [66, 110, 111]. The further addition of the MS module to GC-MS instrument leads to GC-MS/MS [107]. Superior performance is achieved by single and triple quadrupole modes.

The significant vapor pressure of the analyte should be between 30 and 300°C for analysis with GC [107]. Yinon has shown the use of GC-MS with a temperature-programmed injector for some thermally labile molecules [9]. Higher injection port temperature is needed for explosives with low-volatility (*e.g.*, TNT), while much lower temperatures are required for those with high thermal lability (*e.g.*, nitroglycerine (NG)) [109]. Even RDX and PETN have low volatility, they were successfully and rapidly detected by GCxGC TOF MS with a good resolution in a study [112]. This study had a qualitative aspect. The development of an SPME-GC×GC-TOF-MS method to identify the signature of volatile explosives was the target of the work. The samples were analyzed both in solution (liquid injection) and in solid-state (SPME. The injector temperature was 200°C. The GCxGC configuration was made of semi-polar BPX-50 (6 m × 0.25 mm ID × 0.25 μm df) column as ^{1}D and a BPX-5 (0.6 m × 0.25 mm ID × 0.25 μm df) column as ^{2}D. 5 min GC×GC analysis was a suitable way to overcome the co-elution of the components of the volatile signature in the 1D GC analyses. This method allowed a preliminary identification for the nitro explosive compounds in the samples, with an excellent separation performance. The MS library match was within acceptable limits for the analytes, and the retention times were reproducible. The GC×GC conditions were optimized for fast separation, using a short column length, a high carrier gas flow rate, and a fast temperature ramp. Different explosive compounds (RDX, PETN and TNT) were detected as well as taggant molecules and plasticizers. The low elution temperature and the low LOD in fast GC×GC analysis, has also lead to the detection of less volatile compounds.

Another limitation for standard GC-MS instruments is the obligation of low injection volume for samples in split/splitless injectors. In routine analysis, this injector can function with a typical sample volume of 1–2 μL, with a maximum volume of 5 μL. Large volume injection (LVI)-GC technique is superior for semi-volatile and thermally labile compounds and may be utilized to achieve a higher sensitivity through introducing an efficient volume as; tens of microliters or to even hundreds of microliters of the sample [109]. However, reaching large volumes of solvent vapor from a sample to the MS system is one of the most critical problems in LVI which should be prevented. Various injectors and techniques have been designed to fight with this problem. Among these, the programmed temperature vaporizing (PTV)-large volume injection (LVI) injector now is the most used one, which is followed by on-column LVI technique. The on-column LVI is based on a deactivated uncoated short column as a retention gap, which is connected to an early solvent vapor exit before the column of GC. While introducing the sample, the solvent vapor exit remains open to remove it from the system, but the concurrent loss of volatiles causes difficulty in control. The compatibility of the technique with MS is limited because of the high load of solvent that reaches to the MS detector. To solve this problem, PTV-LVI-GC-MS has been used in splitless mode in the literature, utilizing ionization modes as EI and NCI [34, 109]. The PTV-LVI is based on a temperature-controlled function injector, where fast cooling or heating is possible during injection to eliminate the solvent overload at the injector [109].

For thermolabile analytes, the smaller internal volume of PTV inlet in splitless mode is more advantageous since sample vapors in a hot vaporizing chamber will remain shorter and thermal decomposition is anticipated to decrease. GC inlet temperature is another essential parameter to be able to avoid thermal decomposition. In Fig. (**1**), the mean relative responses for thermolabile analytes are given by Kirchner *et al.* [34] for various inlet temperatures in EI mode. The best responses for TNT, pentrite, and RDX were obtained at 150 and 175°C, respectively. The increase of relative responses of RDX was visible when the temperature was raised from 125°C to 150 and 175°C because of better vaporization. At these temperatures, pentrite begins to decompose slightly. At the higher inlet temperatures (>175°C) thermal decomposition caused a decrease in response to the analyzed compounds. RSDs increase with increasing temperatures; the lowest values were reported at 150 and 175°C. To prevent polar organic analytes from being adsorbed on active sites on the surface of inlet liner, retention gap and/or analytical column, analyte protectants are added to the sample and to the standard solution to be adsorbed on the active sites in order to decrease the number of active sites inside the injector. This results in the decrease in polar analytes adsorbed, higher responses, and elimination in peak tailing. GC-MS analyses were carried out with analyte protectants using PTV in solvent vent mode at an initial temperature of 40°C for 0.1 min, vent flow of 50 mLmin^{-1} for 0.1 min, temperature ramp of 300°C min^{-1} to 175°C, purge time of 1.13 min and purge flow of 100 mLmin^{-1}. The oven was adjusted to 50°C (1.13 min) and the ramp was applied as 90°C min^{-1} up to 220°C. 3 analyte protectants solution which are 3-ethoxy-1,2-propanediol and D-sorbitol, L-gulonic acid χ-lactone were used at approximate concentrations of 40, 4, 4 mgmL^{-1}, respectively. The solvent for these solutions was acetonitrile: water (7:3, v/v). 20 µL of analyte protectants solution was added to 1 mL sample. LODs in ng/mL levels were obtained as a result of low temperature and use of PTV-LVI. Especially the high LOD for RDX was decreased significantly, using analyte protectants.

The PTV inlet introduced a novel "solvent-vent split mode" injection, where a high volume of sample can be injected into a packed liner inlet beginning at a low temperature with the split valve open, where the solvent overload can be removed with a high-low temperature change before the injection is switched into a standard split or splitless mode for heating [109]. Various packing materials and inserts gives the opportunity of determination in samples with dirty matrix . Improved detection limits are obtained through conventional LVI analysis carried out using solvent-vent mode PTV or on-column systems. But conventional LVI analysis has limited performance in multi-analyte detection. Moreover, problems can be experienced with some thermolabile explosives.

In 2015, Marder *et al.* [109] developed another novel technique in GC-MS, which can achieve trace analysis as 4 orders of magnitude, over a broad range of molecular volatilities including volatile analytes, to detect challenging explosive compounds in a larger scale. The unique double-column configuration setup removes the solvent overload using a flame detector before the analytes arrive at MS, along with the exact timing of carrier-gas flows and the heating program. Samples undergo three steps after combining a commercial thermal desorption system (TDS) unit to GC-MS, along with the extraction of liquid samples using thermal desorption sorbent tubes; (1) thermal desorption (analyte adsorption), (2) cryo-trapping of the analytes along with venting the solvent, controlled heating to transfer the trapped analytes to GC-MS. In spite of using the solvent-vent split mode injection at the PTV inlet, it was adjusted to a constant splitless mode during the entire injection step. This mode helps to introduce both volatile and semi-volatile compounds to the system. Thus, to prevent the solvent reaching the MS detector, a backup column was used and programmed with a high carrier gas flow to rescue the analytes from the intensive amount of solvent *via* elution through the column, then the excess solvent was vented through a second selective detector for solvent waste during the run. The continuous separation of the solvent from the compounds through the entire chromatographic column prevented the loss of volatile analytes along with the solvent. The explosives were extracted using 5 mL methyl tert-butyl ether (MTBE):acetone (4: 1, v/v) solution.

Two parallel columns connected to a separate injection port each were combined through splitters 1 and 2, as shown in Fig. (**3**). Separation of the analytes was performed through column A (2 m DB1 precolumn, 0.32 mm x 0.1 μm) connected with a straight connector to Column B (15 m DB5-MS column, 0.25 mm x 1 μm) (Column B). As seen from Fig. (**3**), both columns were integrated through splitter 1. To overcome the problem of vacuum overload, a back inlet was connected to a second backup column (B). Both columns A and B were connected to MS, and Flame Photometric Detector (FPD) through splitter 2. Here, FPD only to serves as a waste collecting point. So, it was operated at 250°C without additional gas flow. For the analysis, 20 µL samples were injected at a slow and constant rate of $2\mu Ls^{-1}$ using a 100 mL syringe.

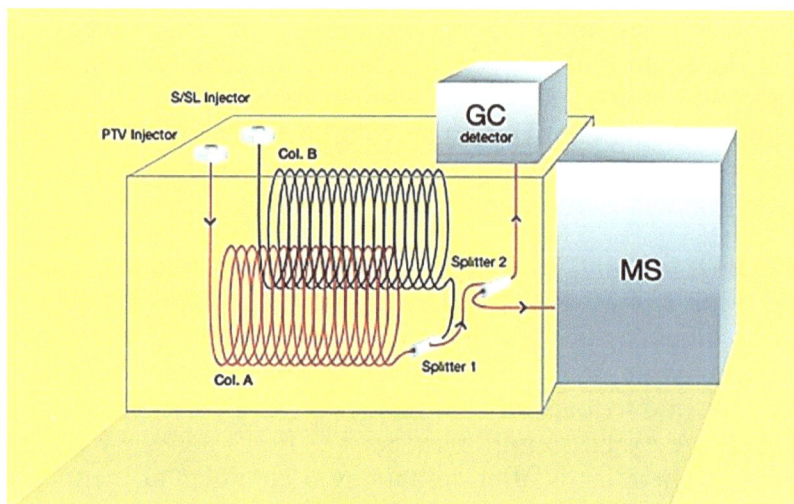

Fig. (3). Instrumental configuration of the splitless-PTV-LVI-GC-MS method. The path of the analytes is demonstrated in red. Used with permission of Royal Society of Chemistry, from Ref [109], permission conveyed through Copyright Clearance Center, Inc.].

Very high sensitivities are obtained with this method for military and improvised explosives. LOD and LOQ values were 0.1 $pg\mu L^{-1}$ (2 pg on column) with a linear dynamic range of at least 2 orders of magnitude. LOD was 0.5 ngg^{-1} in spiked soil samples. The detection of the challenging compound RDX was enabled using the same method configuration, but after changing the packed liner to baffled and increasing the column flow (Method 2). The chromatograms of the explosives in their 0.1 $pg\mu L^{-1}$ standard mixture solution, by-products and taggants after 20 µL injection are given in Fig. (**4**).

Chemical identification methods with GC-MS usually depend on the comparison of the fragmentation patterns of unknown compounds with those in the GC-MS library [113]. Even though an MS library search usually matches the compounds with correct chemical identities, there may be mistakes in chemical assignments of structurally similar compounds. Detectors such as absorption spectrometers have been used for the analysis of volatile organic compounds and can provide complementary absorption data. These techniques can be hyphenated with tandem MS for more reliable discrimination of structurally similar volatile organic compounds. For example; the isomer selectivity of vacuum ultraviolet absorption spectroscopy without destroying the analytes contributes to the mass selectivity of vacuum UV (VUV)-MS in the powerful identification of the compounds in GC-MS analyses. Orthogonal VUV and MS library searches present a more powerful characterization of volatile organic analytes in complex matrices. This technique also seems worth to try for the volatile explosive compounds. Furthermore, this technique can be combined with the above mentioned PTV-LVI technique above [109], to enlarge the scale of detectable compounds and to provide a more precise identification.

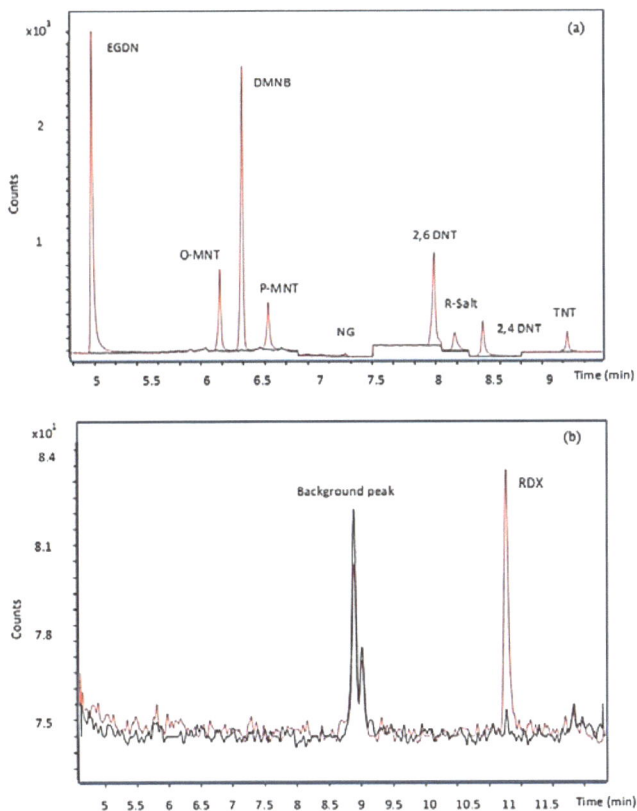

Fig. (4). Chromatogram of 0.1 pg L^{-1} standard mixture solution of explosives, by-products and taggants after 20 L injection f in MTBE: acetone (4: 1 v/v) (red) and a blank sample (black) in method 1 (EGDN-TNT) (a) and method 2 (RDX) (b). Used with permission of Royal Society of Chemistry, from Ref [109], permission conveyed through Copyright Clearance Center, Inc.].

Determination of ammonia from ammonium nitrate (AN) is one of the important issues from the point of forensics and security, as AN is a common component of homemade explosives (HMEs) [72]. Ammonia vapor in the headspace of AN salt arises from its dissociation producing forming ammonia and nitric acid. Although there are many commercial sensors for detecting ammonia, detection of trace levels is a challenging issue from the point of its small mass, high volatility and polarity. The detection is hard, particularly in the presence of other volatile organic analytes. Low LODs are possible with SPME, but very volatile analytes as ammonia generally do not retain well to the fiber coatings of SPME. However, Brown *et al.* demonstrated that an alkyl chloroformate-derivatized SPME fiber provides a good extraction and identification of ammonia using GC-MS [73]. Converting the ammonia to a higher molecular weight derivative, allows a favorable partitioning of ammonia on the fiber coating, providing better GC-MS parameters for this compound. One of the superiorities of this method is it's use in AN-based homemade explosives detection, which detects derivatized ammonia (as butyl carbamate) preventing the interaction with other headspace components, but also detecting the other signature components in the vapor in one run. The detection of ammonia in the headspace of solid ammonium nitrate and ammonium nitrate-based home-made explosive mixtures using an internal standard, along with the fuel components in headspace is demonstrated. Externally sampled internal standard (ESIS) may be helpful in quantification in analyses done with SPME, which is regarded as a semi-quantitative technique [114]. Although the above-mentioned advantages, the use of ESIS and IS can make the analysis complex and tedious. Instead of SPME, at 2019, Katilie *et al.* [72] trapped and derivatized ammonia directly on a cryo-cooled GC inlet, for simplification and elimination of several steps and the need for an internal standard for quantitation in the previous SPME approach. Ammonia was derivatized on line with butyl chloroformate on a cryo-cooled glass wool in GC inlet liner to produce butyl carbamate, a compound that is retained by GC columns and compatible with standard GC-MS analysis. A successful quantification of ammonia has been achieved in the vapor produced from a permeation tube and from solid ammonium nitrate without loss of sensitivity in SPME The success of the method was demonstrated for two homemade explosives Ammonium nitrate-aluminum and ammonium nitrate-petroleum jelly.

With commercially available sensors, poor selectivity and sensitivity are experienced with LODs at ppm level or higher [115]. Ammonia detection is also a challenging issue with conventional laboratory techniques, as GC-MS, where incompatibility with typical gas chromatographic stationary phases results in poor retention and peak shape [72]. Since the instrumentation techniques as ion mobility spectroscopy, ion chromatography, and capillary electrophoresis, are incompatible with direct vapor detection, they require additional sampling and sample preparation steps for the detection of ammonia vapor where the efficiency, and the sensitivity decreases. There are state-of-the-art chemical sensors providing good selectivity/specificity for ammonia but yet can't detect trace amounts. However, the above method can successfully detect ammonia with GC-MS, using a derivatizing agent collected on a cryo-cooled inlet of GC, especially in the presence of interferents.

Analysis with LC-MS & LC-MS/MS

Though new improvements have been emerged in GC-MS analysis of explosives, there are still disadvantages of GC one of which is the adsorption of polar explosives in the chromatographic system, causing a decrease in response and peak tailing [34]. Also at high temperatures typically used in GC, there is a risk of possible decomposition of the explosive analytes due to thermal lability, however, LC-MS and LC-MS/MS provide superiority over GC-MS and GC-MS/MS in the analysis relatively polar molecules and heat labile compounds which may degrade at these temperatures.

Nitroamines and nitroesters are preferred to be analyzed using LC/MS instead of GC/MS, because of their low vapour pressure and thermal lability [78]. LC-MS/MS is also superior to LC-MS in selectivity and sensitivity in complex matrices and multi-analyte determination in one run. False negatives are partially eliminated using the MS-MS technique [30]. However, explosive analysis methods developed using LC-MS/MS are still limited.

The most commonly used ionization methods for explosive analysis are atmospheric pressure ion sources (API) such as electrospray ionization (ESI) or atmospheric pressure chemical ionization (APCI) combined with an HPLC or UPLC system and mainly used for analytes with low molecular mass. ESI is suitable for the determination of hexogen (RDX), octogen (HMX) and nitro pentaerythritol (PETN), because of the high sensitivity obtained [58]. A method carried out using stir bar extraction before analysis with APCI(-) LC/MS in literature could detect lower concentrations than SPE, required shorter time and less chemicals. Among DNB, TNB, DNT, TNT, and PETN, PETN could be determined with lower precision and the detection limits are higher for some common nitroaromatic explosives such as trinitrotoluene (TNT), dinitrotoluene (DNT) or trinitrobenzene (TNB). ESI is claimed to suffer of lack of reliability in the analysis of these explosives because of the matrix effects that can arise during the analysis of real samples. APCI is recognized to be very reliable, producing less variation of matrix effects. As a result, the composition of the organic solvent, pH, and ionic strength of the samples can be modified with very little variations in signal intensity.

Mass spectrometry of nitro compounds provide information on their thermodynamic and energetic data *e.g.* fragmentation pathways, ion-neutral reactions, proton transfer, and electron capture [78]. Optimizing compound/source specific parameters in the method is important in obtaining proper ion formation. Many organic compounds have low ionization energies (5– 15 eV), so easily get rid of an electron. Fragmentation and rearrangement of the structure occur with the excess energy; because of this, the ionizing energy is decreased, production of the molecular ion is possibly favored, and fewer fragments are produced. The ionization potential of compounds, help to predict if the ions will form and the value to which the applied voltage can be adjusted. The analytes which bear nitroaromatic/amine/ester in their structure, have relatively high electron affinities and expected to produce negative ions in the negative mode of mass spectrometry. Sometimes, the analytes are identified in the form of adducts at higher sensitivity, after the addition of the ionic modifiers.

To improve the sensitivity in ESI-LC/MS, the sprayer voltage and position are optimized. The cone voltage (also known as declustering potential or orifice voltage) may be optimized to decrease cluster (or adduct) formation, which can be detrimental due to their stability for fragmentation. As the pH increases, the proton is expected to dissociate easily. In explosives with high gas-phase acidity numbers such as RDX and HMX, proton–anion dissociation is not expected. RDX, HMX and PETN form adducts with the additives. Furthermore, the steric effects can affect the ionization pattern.

Use of Triple Quadrupole MS and hyphenated instruments with ToF-MS, Ion Trap and Orbitrap

Negative photoionization is generally proposed for organic explosives in MS analysis. Liquid chromatography/negative ion atmospheric pressure photoionization mass spectrometry (LC/ NI-APPI-MS) was used in full scan mode, as a novel and highly sensitive method for the analysis of DNB, DNT, TNB, TNT, HMX and RDX, where LOQ values were two orders of magnitude lower compared to APCI-LC/MS [116]. ToF-MS technology was used. The MS parameters were as follows: gas 1 or nebulizer gas: 30 arbitrary units (augas 2 auxiliary gas: 30 au; curtain gas: 40 au, IS: ion spray voltage, -1500 V; temperature: 350°C, declustering potential:-25 V; focusing potential: -80 V, declustering potential 2:-12 V; and collision gas pressure 3 au. The method was reported to display good reproducibility and high intraday/interday precision.

ESI(-) LC-MS/MS was used in surface water analysis, and the method had detected the nitroesters RDX, HMX and PETN as acetate adducts and TNT as a quasi-molecular ion [117]. SPE and direct injection analysis were compared in the trace analysis of explosives such as TNT and its mono and diamino metabolites, HMX, RDX, nitroglycerin, and PETN in lake water and tributaries. A novel SPE adsorbent material and a phenyl based stationary phase HPLC column were used. ESI (+) mode was applied for diamino metabolites of TNT and (-) mode was applied for all other compounds. The method was validated and the LOQ values in preconcentrated lake water samples were between 0.03 - 1 ng/mL and 0.1 - 5 ng/mL in river water. Direct injection analysis revealed comparable results to preconcentrated water samples for the most persistent explosives. Analysis of lake water samples collected at different depths and main tributaries showed the presence of HMX, RDX, and PETN at concentrations between 0.1 - 0.4 ng/mL and 0.1 - 0.9 ng/mL of the same compounds. Since no significant difference was observed between direct injection of samples and SPE method, direct injection is regarded as suitable and a fast technique for water samples with cleaner matrices.

In one of the LC-MS/MS studies on explosives, Şen *et al.* [30], have developed a rapid analysis method for HMX (cyclotetramethylenetetranitramine), RDX (siklotrimethylene-trinitramine), TNT (2-methyl-1,3,5-trinitrobenzene), picric acid (2,4,6-trinitrofenol), tetryl (2,4,6-trinitro phenyl-methylnitramine) and PETN (Pentaerythritol tetranitrate), and demonstrated its success in post-blast debris in real cases. Inertsil ODS-4, 100×2.1 mm, 3μm was used as the stationary phase at 40°C. 5 mM aqueous ammonium acetate (A) and 5 mM methanolic ammonium acetate (B) were used as the components of the mobile phase. ESI needle voltage was set to 3.5 kV, the pressure of the nebulizer gas to 3 L/min, and drying gas to 15 L/min. The peaks were eluted in 8 minutes with mobile phase A:B (5:95, v/v).

The cotton used in the sampling processes was obtained from a regional supermarket and was pre-cleaned by washing with methanol. Post-blast evidences sent to the laboratory in a real case, swabs, and cotton samples used for swab were extracted with methanol. The extract solvent was concentrated through evaporation. Water was added to the resulting concentrated samples and analyzed. The most suitable solvent composition for the samples prepared for analysis was determined as 50/50 methanol: water. The chromatograms are given in Fig. (5).

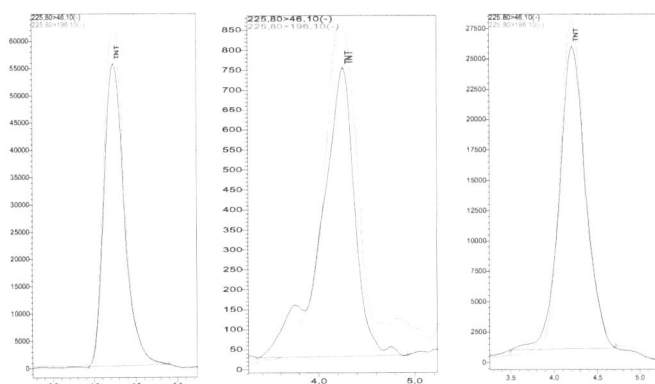

Fig. (5). Chromatograms of scarves, metal particles and swabs (respectively) [Reprint from Ref. 30].

According to the results of the analysis, it is seen that explosive material residues are higher in fabric pieces (especially scarves) and pieces of ceiling trims compared to metal particles and swabs (Fig. **6**).

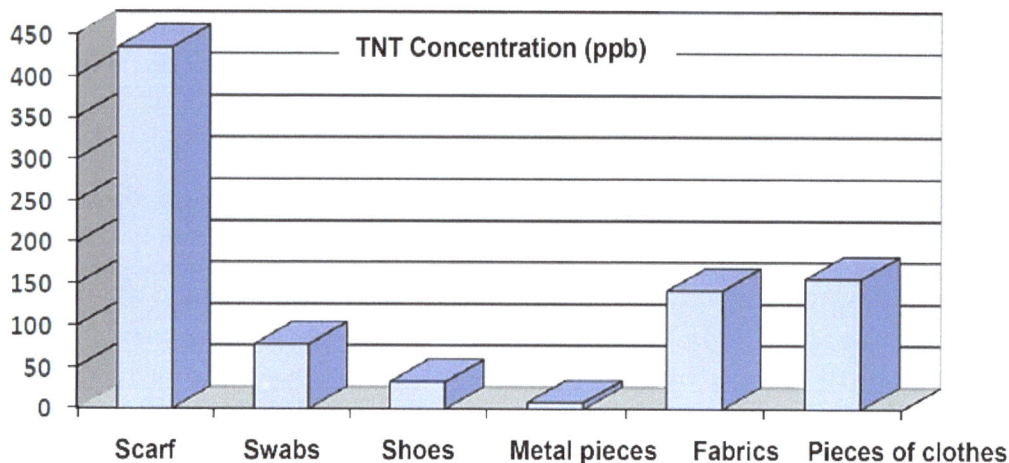

Fig. (6). TNT concentrations in ppb, obtained as a result of the analysis of the evidence in the real event [Reprint from Ref. 30].

The amount of explosive residue that may be present in the findings exposed to the same extent to the explosion will vary according to the thermal conductivity and porosity of the evidence. Explosive substances easily decompose by heat due to their chemical properties or evaporate easily, even if they do not decompose. The possibility of the presence of residues of explosives on metallic materials that have high thermal conductivity is higher than that of fabrics, *etc.* which have low thermal conductivity.

Many established methods in the literature are focused on water and soil matrices. Since it is a problematic matrix, some of the existing LC-MS/MS methods in the soil are directed to qualitative purposes, tedious and time-consuming, some have longer times of chromatographic analysis, or higher amounts of solvent/sample ratios and some of them lack recovery or some lack validation. To detect trace concentrations of explosives before their decomposition, rapid, sensitive, less laborious, reliable, easy, and economical techniques are needed. Simultaneous analysis methods of explosive molecules with minimum solvent use is desired.

Triple quadrupole LC-MS/MS instruments provide very satisfactory results and chromatograms [69, 71, 118]. Triple quadrupole MS, combined with HPLC or UPLC provides very low LODs for explosive residues in multi-component analyses [118]. ESI is very effective for nitramines and nitro esters, when used in negative ionization mode, especially for nitrate, chloride, formate and acetate adducts. However, the sensitivity decreases for nitroaromatics. APCI(-) ionizes the adducts of nitramines, nitrate esters and nitroaromatic compounds due to their strong electronic affinity and low basicity [70]. Fast and simple LC-MS/MS methods are developed using ESI and APCI, in the soil matrix, which is the most common available samples encountered in the crime scenes. A green method for the determination of PETN, RDX, and HMX in 1 g soil was developed using a one-step 2.00 mL acetonitrile extraction and the validation was performed by Anilanmert *et al.* [69]. After 3 min vortex, ultrasonic-assisted extraction (UAE), which is a very effective way of fast and easy sample preparation, was applied for 15 min. After centrifugation, decantation of supernatant and evaporation under N_2, the residue was dissolved in 500 µL acetonitrile, filtered and analyzed. Mean recoveries were in the range of 76.52-84.77%. Since the solvent use was minimum, the method was environmentally friendly, fast and sensitive with low LOD (3.4-8.5 ngg^{-1}) and LOQ values (6.0-10.0 ngg^{-1}). Matrix effect was investigated and 5 soil matrices were compared, as seen in Fig. (**7**).

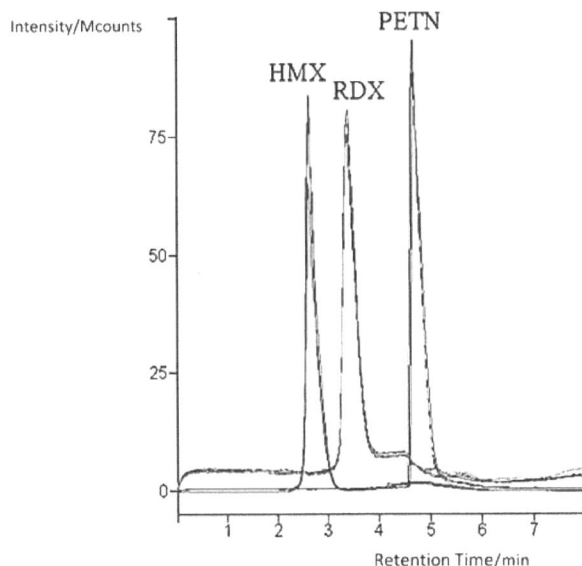

Fig. (7). HMX, RDX and PETN in 5 different soil matrices (overlaid) (500.0 ng/g) ["Reprinted (adapted) with permission from "Anilanmert, B.; Aydin, M.; Apak, R.; Avci, G.Y.; Cengiz, S. A fast liquid chromatography tandem mass spectrometric analysis of PETN (pentaerythritol tetranitrate), RDX (3, 5-trinitro-1, 3, 5- triazacyclohexane) and HMX (octahydro-1, 3, 5, 7-tetranitro-1, 3, 5, 7-tetrazocine) in soil, utilizing a simple ultrasonic-assisted extraction with minimum solvent. *Anal. Sci.*, 2016, 32(6), 611-616).". Copyright (2016) Japan Society for Analytical Chemistry." Ref. 69].

The method was successfully used in artificial explosion and real samples. In 2010, a bomb attack took place in Taksim Square, Istanbul [119]. The target of the terrorists was the flying squad. A man had walked to the area of the police on duty at Taksim Square and pulled the pin of the bomb. 15 police officer and 17 citizens were wounded. With the method developed, very low concentrations could be detected even after several times of wash of the ground of the post-explosion crime scene, where RDX was found in high amounts on the concrete and soil samples collected from the side of the wall and grass. HMX was also detected in trace amounts. Post-explosion crime scene at Taksim Square at 2010 can be seen from the website of NationalTurk news [119]. The chromatograms obtained from real samples in Taksim and model samples obtained from Machinery and Chemical Industry Institution (Ankara) are given in Fig. (8).

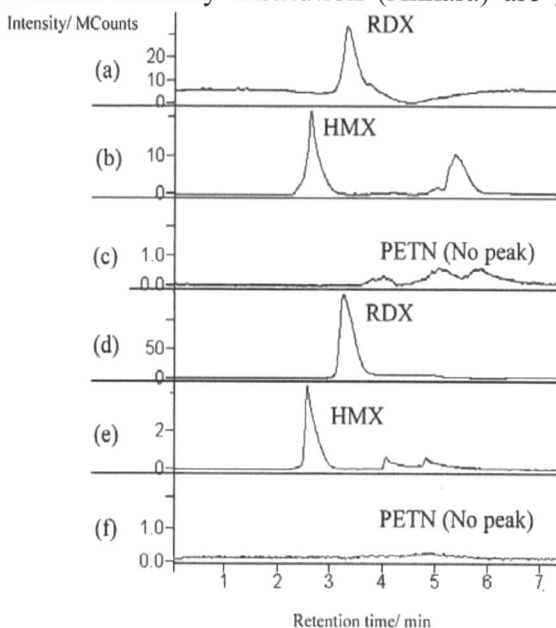

Fig. (8). a, b, c: model sample obtained after an artificial explosion at Machinery and Chemical Industry Institution (Ankara). d, e, f: Taksim Square Austrian source *"SCH Portfire A4'* formerly used by. Devrimci Karargah, DHKP-C & PKK ["Reprinted (adapted) with permission from "Anilanmert, B.; Aydin, M.; Apak, R.; Avci, G.Y.; Cengiz, S. A fast liquid chromatography tandem mass spectrometric analysis of PETN (pentaerythritol tetranitrate), RDX (3, 5-trinitro-1, 3, 5- triazacyclohexane) and HMX (octahydro-1, 3, 5, 7-tetranitro-1, 3, 5, 7-tetrazocine) in soil, utilizing a simple ultrasonic-assisted extraction with minimum solvent. *Anal. Sci.*, 2016, 32(6), 611-616)."". Copyright (2016) Japan Society for Analytical Chemistry. Ref. [69]].

Devrimci Karargah, DHKP-C & PKK [Reprint from Ref 69]. Another sample preparation and the chromatographic method was developed for TNT, RDX and HMX determination in soil, using a different extraction and chromatographic elution procedure [70]. This time a more effective extraction procedure was developed with acetone use, for the explosives. Higher recoveries between 96.9 - 100.4% were obtained. LOD and LOQ values obtained from the analysis of the spiked soils were 4.3 - 6.8 ngg^{-1} and 7.00 - 10.0 ngg^{-1} for RDX and HMX, and 18.9 and 38.0 ngg^{-1} for TNT. An LC–APCI–MS/MS screening method was also developed to determine the trace amounts of TNT, RDX, HMX, PETN, TETRYL, picric acid, 2,6-DNT and TMETN in the soil after the explosion [71]. In the literature, studies regarding multiple explosive analyses with LC–MS/MS in soil are limited. To determine TMETN in the presence of RDX, HMX, PA, TNT, TETRYL 2,6-DNT and PETN in one run was challenging and a different technique was used to achieve this separation. Since it was hard to observe the specific primary and secondary ions for all the above-mentioned explosives in one run, no satisfactory simultaneous analysis method existed in literature for these explosives in soil. An extraordinary gradient approach in MS/MS gas temperatures, successfully provided a selective LC-MS/MS elution after a single-step extraction with acidified acetonitrile. Recoveries were between 93.01 - 104.20%, with LOD and LOQ values between 8.9 – 161.2 and 13.2 – 241.5 ngg^{-1}. The elution time did not exceed 10 minutes.

In a study in 2016 [58], stir bar sorptive extraction was used followed by liquid desorption and LC–MS/MS analysis. 8 analytes from different chemical classes including nitro-aromatics, amino-nitro-aromatics and nitric ester compounds, have been monitored. The method was developed and optimized using a statistical design of experiment approach. The optimization strategy is based on the use of full factorial and Doehlert designs for testing a total of 10 parameters for a complete study of the experimental space. During the chromatographic elution, a HypersilGold PFP (pentafluorophenyl) column (2.1 mm × 100 mm, 1.9 μm particle size) at 35∘C was used. The mobile phases A and B were the aqueous solution of 1 mM ammonium formate and MeOH. The composition of the mobile phase was 25% B initially. Then, a ramp of 1 minutes was applied to reach 30% B. A second ramp was applied for 10 min from 30% B to 55% B at 10 min. Another ramp from 55% B to 80% B was performed for 12 min. Then the composition was changed to 95% B at 13.1 min. Then the program was switched to the initial composition at 13.5 min for equilibrating the column for 3.5 min.

Ionization source conditions were as follows: corona discharge voltage 40 V; vaporizer temperature 350∘C; sheath gas pressure 40 psi; sweep gas pressure 10 psi; auxiliary gas pressure 15 psi; capillary temperature 220∘C; skimmer offset 0V. Collision energies and tube lens voltages were optimized for each analyte. Ranges of LOD and LOQ values were 0.008-0.7 ng/mL, and LOQ 0.01-2.1 ng/mL, in order. In another study, London wastewater was analyzed using APCI(-) LC/MS NB, TNB, TNT, DNT, HMX, RDX and PETN were identified, except NB [81]. The study has the characteristic of a general method for trace analysis for multiple classes of explosives: nitroaromatics, nitroamines, nitroesters and peroxides. 14 analytes were detected at $ngmL^{-1}$-μgmL^{-1} level, among the 17 tested and the recoveries ranged from 77% to 124%. Nowadays, many hybrid techniques are on the rise, combining powerful mass analyzers as TOF (time-of-flight), Orbitrap, and ICR (ion cyclotron resonance), with quadrupole or ion trap instruments [78]. LC-QTOF instrumentation has the same configuration of a quadrupole mass spectrometer, with the third quadrupole regarded as a drifting tube. LC-QTOF/MS technique presents a better identification and information of analytes and accurate mass even at sub-ppm levels, because of high resolution [80]. Analyte identification chance increases because of the possible number of candidate compounds are restricted through these features. The hyphenation of a quadrupole and a collision cell with ToF analyser, provides a better selectivity *via* enabling the fragmentation of pre- selected ions and the identification of compounds using their product ion spectra. Thus, mass determination of parent and daughter ions is performed accurately. Standards are individually analyzed from the point of retention time, observed spectral molecular weight, and characteristic fragmentation pattern. Then these data are recorded in the library for comparison for the identification of analytes in samples. High-resolution MS techniques like LC-Orbitrap MS gives the chance to identify unknown compounds even there are no available reference standards. HPLC-(PDA)-LTQ Orbitrap FTMS, is a recently developed instrument, which performs the acquisition of full scan MSn (n-stage fragmentations, $n = 1 - n$) spectra with an LTQ detector at high speed and/or with the Orbitrap having ultra- high mass resolution and high mass accuracy [120].

The power of a mass analyzer to resolve the peaks in a mass spectrum is regarded as resolution power (RP). Peaks are decided as resolved if the valley between them is equal to or less than 10% of the intensity of the weaker peak when using magnetic or ion cyclotron resonance (ICR) instruments. This limit is 50% for quadrupoles, ion trap, TOF, *etc* [120]. Mass accuracy is mostly related to the stability and resolution of the instrument. Only a high resolution can lead to high accuracy. Instruments with high accuracy can be used for the determination of the elemental composition. According to the classification of Holcapek *et al.* [121] most quadrupole (Q) and ion trap (IT) instruments have low-RP, TOF-based analyzers have high-RP, and two Fourier transform (FT) mass analyzers Orbitrap and ICR, have ultra-high RP. High resolution mass spectrometric (HRMS) techniques as Orbitrap, QTOF, *etc*. are particularly used in identification of new analytes and enantiomers. The resolution of spherical ITs and especially linear ion trap (LIT) analyzers is slightly better than that of quadrupole analyzer, but a decrease in sensitivity occurs as the resolution increases [120]. A reliable determination of elemental formula cannot be performed because of unsufficient accuracy in low-resolution mass analyzers. From the point of accuracy, the Q rods with ideal hyperbolic profiles provide a higher resolution than the regular round Q rods, while the acquisition speed is lower. ICR, Orbitrap, and TOF are the instruments with the best RP and mass accuracy, respectively. When FT analyzers are used, the image currents should be recorded in adequate repeats to obtain the best resolution and accuracy. TOF analyzers have the highest scan speed and high selectivity even with overlapping peaks and a matrix with high complexity. TOF has the superiority of estimating the empirical molecular formula using the accurate molecular mass and isotope peak pattern [122]. Structural information can also be received comparing the accurate masses of the ionic fragments and neutral losses with the postulated fragment formulas. A comparison of the mass analyzers are displayed in Table **1**.

Table 1. Comparison of some old and relatively new mass analyzers.

Type of Mass Analyzers	Abilities and Uses	
ICR	-ultra-high RP -better RP and mass accuracy – than Orbitrap	-high priced
Orbitrap	-ultra-high RP -suitable for the identification of new analytes and enantiomers -better RP and mass accuracy than ToF	-moderate to high priced
ToF	-high RP -high accuracy -suitable for identification of new analytes and enantiomers -highest scan speed and high selectivity even with overlapping peaks -can estimate the empirical molecular formula using accurate molecular mass and isotope peak pattern -structural information can be received comparing the accurate masses of the ionic fragments and neutral losses with the postulated fragment formulas.	-sensitivity is not high as triple Q -lower dynamic range of this type of analyzer it will not be as easy as with triple Q.
Quadrupole	-Q rods with ideal hyperbolic profiles provide a higher resolution than the regular round Q rods -higher sensitivity and higher linear range is possible with triple Q	-low RP -a reliable determination of elemental formula cannot be performed because of unsufficient accuracy
IT	-RP and sensitivity are slightly better than quadrupole, but sensitivity decreases as the resolution increases	-low RP -a reliable determination of elemental formula cannot be performed because of unsufficient accuracy
LIT	-RP is slightly better than quadrupole	-sensitivity decreases

Even though QTOF/MS acquires data rapidly (500 scans/s) compared to Orbitrap (10 scans/s), its resolving ability is lower(~60,000@FWHM) than Orbitrap (~100,000@FWHM) or an ICR (1,000,000@FWHM) [78]. In conclusion, a high-resolution MS instrument significantly eliminates the probability of false positives, arising from the sample. A high resolving power is not adequate for discrimination between isotopic molecules; thus, high mass accuracy instrument is necessary for trace analysis. The highest mass accuracy among ICR, Orbitrap and QToF) belongs to ICR (0.1 – 1 ppm), and this is followed by Orbitrap (0.5 – 1 ppm) and QToF (3 – 5 ppm). Hybrid instruments are used to overcome problems related to isotopic confirmation.

In 2014, Xu *et al.* [12], developed a highly selective screening of a post-blast explosive sample, having a complicated matrix. The analysis method was developed using swabbing. Swabs obtained from post-blast debris were extracted using 10 mL methanol. The extract was evaporated up to 0.5 – 1.0 mL and analyzed with HPLC–(PDA)–APCI negative ionization-LTQ MS2/Orbitrap FTMS. The method was developed for qualitative purposes. This new hybrid instrument, which combines an IT and Orbitrap with HPLC, was able to identify isomers of DNB and DNT, with LODs at pg level and mass accuracy within 1 ppm.

DIRECT ANALYSIS TECHNIQUES

Ambient MS (DART-MS, DESI-MS, and Similar Techniques) and Hyphenation of Ion Mobility and IR Techniques with MS

The conventional direct analysis technique is generally IMS. Ion mobility spectrometers (IMSs) are fast and sensitive instruments utilized also for the determination of explosives [123]. In traditional ion mobility spectrometry (IMS), the sample is introduced into a reaction region where ions are formed [124]. A narrow pulse of ions is injected into a drift region and move along this drift tube with the help of an applied electric field (1-500 V/cm). In the drift region, ions are separated according to their low-field mobility, which is related with their charge and size. The IMS can work at atmospheric pressure and room temperature.

In IMS, the ions are separated by gas-phase viscosity instead of molecular weight as in MS technique [125]. The accuracy is significantly lower than MS for complicated samples or simultaneous multi-analyte screening. IMS is economical, is suitable to be used as a handheld analyzer for rapid, preliminary detection of explosives. IMS use is widespread for explosive determination before detonation. It is firm, portable, and can be used in on-site detection [2]. However, its use in the laboratory has certain limits: weak resolution, limited selectivity, and chemical interference, therefore these analyzers cannot always be applied directly to comprehensive screening analysis. The trend for the development studies for IMS target; miniaturization, substitution of non-radioactive ionization sources, and to increase the performance of the instrument.

Coupling IMS with high speed GC provides detection of a range of explosives at ng levels in seconds [126]. Drift tube ion mobility coupled to MS (DTIMS) is recommended to be used in the discrimination of home-made explosives from industrially manufactured explosives available in the market, and this is a very sensitive tool to eliminate false positives in the forensic cases [78, 127].

Ambient MS has a much wider range of applications and offers much higher sensitivity and selectivity [128]. Ambient pressure laser desorption (APLD) and ambient pressure laser-induced acoustic desorption (AP-LIAD) evaporate the explosives and detect them in the vapor phase. These techniques have the superiority of analysis directly from the surface [123]. Solids, liquids, and gases can be directly analyzed using ambient mass spectrometry and they provide selective and sensitive chemical detection. They also have advantages as real time analysis, minimal sample preparation, and the possibility of chemical imaging.

The fame of direct analysis in real time (DART) has increased in the forensic science world with its sensitive detection. DART is a solvent-free plasma-based ambient ionization technique that has a heated gas stream for analyte desorption and the production of helium (He) meta-stable atoms to initiate Penning ionization of atmospheric water for gas-phase analyte ionization [128, 129]. DART, together with DESI, forms the pair of the pioneering ambient MS techniques. DART is a solvent-free technique performed generally on a heated helium gas stream which combines several ionization mechanisms. Ionization is "promoted" by glow discharge that produces electronically excited He* atoms in a helium gas flow. Next, the Penning ionization of atmospheric gases or water present on humid surfaces is promoted by He*, forming a series of ions such as H_3O^+ and O_2^-. Subsequently, *via* atmospheric pressure chemical ionization (APCI), the analyte ions are produced. The thermal desorption nature of gas-phase introduction by DART leads to difficulties for the analysis of low volatility inorganic analytes, which require elevated temperatures for efficient desorption [129]. Typical heater components of DART ion sources can generate temperatures up to approximately 500 °C to 550 °C, however, the corresponding emitted gas stream temperatures experienced by the analyte may be closer to 250 °C to 300 °C and drastically diminish moving away from the source, cooled by the ambient conditions. DART ion source could be enhanced for higher temperatures, sufficient for refractory inorganic salt desorption, however, a thermal degradation of the more volatile and thermally labile organic components in explosive analysis may be a limitation allowing the analysis of only inorganic compounds. Or DART can also be combined with a second heating component, combining the thermal desorption and ionization for better control. A rapid transient heating period would enable volatile compounds to desorb at low temperatures, where inorganic salts could be desorbed *via* elevating the temperature subsequently.

Forbes *et al.* [129] combined a resistive Joule heating thermal desorption (JHTD) with direct analysis in real-time (DART) for the detection of inorganic nitrite, nitrate, chlorate, and perchlorate salts. The heating apparatus produced rapid heating ramps separate from each other with elevated temperatures up to approximately 400 °C s^{-1} and 750 °C, by applying a few amperes of DC current to a nichrome wire and tuning this current. JHTD improved the capabilities of traditional DART-MS also for the detection of inorganic compounds (as ammonium nitrate, calcium ammonium nitrate, potassium chlorate, potassium perchlorate, and potassium nitrite, *etc.*) even at lower nanogram levels, as well as the detection of organic compounds. JHTD filament current and in-source collision-induced dissociation (CID) energy are very important for the response of the instrument. The JHTD-DART presented an efficient desorption of organic and inorganic explosives at their respective optimal temperature for each, followed by rapid cooling, maintaining high throughput. This platform was connected to a ToF analyzer *via* a vapor hydrodynamic-assist interface, which helped in the transportation of the analyte ions toward the ToF orifice. It is suggested by the authors that exchanging the wire material and extending the DC current levels may lead to reach higher temperatures needed for thermal desorption of elemental inorganics (as lead, barium, and antimony components of gunshot residues).

DESI is based on ESI and is a spray-based technique that brings together the useful features of electrospray with the additional benefits of direct surface desorption of the analytes [128]. Typically in DESI, the charged (positively or negatively) droplets generated by ESI is used to bombard the sample surface that is a few millimeters away to desorb and ionize the analyte. As the analyte-containing droplets shrink due to desolvation, the analyte ions are released to the gas phase and travel through the atmospheric pressure interface into the MS inlet. DESI is one of the most important ambient MS techniques used for MS imaging, which is also becoming a powerful tool in forensic chemistry. A recently developed ambient ionization technique, paper spray ionization (PSI) is also based on ESI. In this technique, a high voltage electrospray is generated on a porous triangular piece of paper where the sample and a proper solvent is deposited. PSI provides the advantage of combining extraction and ionization in one step, using solvent-wetted paper cut in a triangle [130]. Different types of PSI techniques were created by replacing the paper by similar surfaces or using matrices as wooden toothpicks [131], swabs [132] or the sample itself, such as in leaf spray [133].

Techniques as DART and DESI present new chances for security screening of explosives since the analytes can be desorbed and ionized directly from the sample surfaces. More number of analytes can be analyzed by DESI than DART [134]. Distant detection techniques are able to detect analytes in the vapors carried over meters, as DESI-MS, which can analyze ambient vapors at 3 m distance, where PETN and TNT can be determined in quite low amounts [130, 134]. Fast detection of explosives in situ from various surfaces were achieved using DESI with a portable MS (LOD <1 ng) [126]. For example RDX was detected on paper. A desorption corona beam ionization source (DCBI) can also be used under ambient conditions with an MS analyzer without any sample preparation. DBDI-MS system has a detection limit range between 0.01 - 0.1 ng for RDX, TNT and PETN [111]. In recent years, other ionization techniques as secondary electrospray ionization (SESI), dielectric barrier discharge ionization (DBDI), selected-ion-flow-tube (SIFT), and proton transfer reaction (PTR), *etc.* were also used for explosives [135]. These systems have detection limits at usually ppt or lower levels. A dual SESI and DBDI source was used with an MS analyzer for real-time fast vapor analysis and used for canine training in the detection of explosives. SESI reaches the analyte vapor stream with an electrospray, and DBDI transports the sample through a plasma or provides a reaction between the analyte and reactive plasma species.

Infrared thermal desorption (IRTD) coupled with DART-MS, is another new hyphenated technique for the detection of both inorganic and organic explosives from wipe collected samples [136]. The system produces separate and rapid heating ramps that thermally desorbs volatile and semivolatile organic explosives at relatively low temperatures, however elevated temperatures can still be used for the desorption of nonvolatile inorganic oxidizer-based explosives. Refractory potassium chlorate and potassium perchlorate oxidizers which are difficult to desorb with traditional moderate-temperature resistance-based thermal desorber, were determined using this system as a model (Fig. **9**). A range of organic and inorganic oxidizer based explosive compounds could be detected at nanogram to sub nanogram sensitivities, where further improvement in sensitivity is limited by the thermal properties of the widespread commercial wipes.

DART was also hyphenated with Orbitrap MS with Raman microscopy for the detailed characterization of explosive molecules [137].

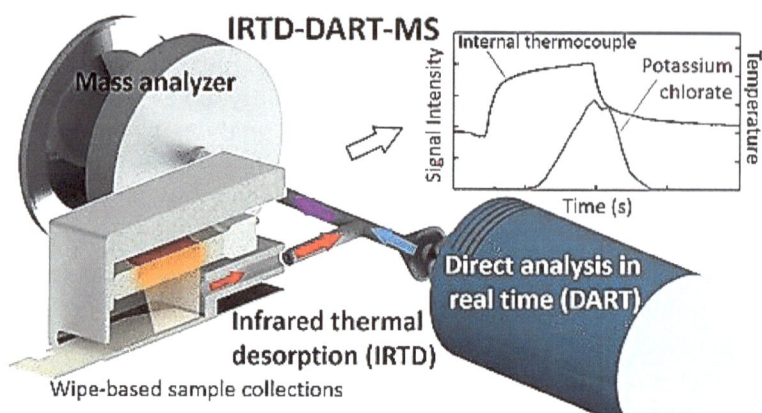

Fig. (9). A block diagram of IRTD-DART-MS [Reprint from Ref. 136].

Ambient MS presents an advantageous platform for the chemical analysis of latent fingerprints [138]. For example, electro-flow focusing ionization (DEFFI) with in-source CID can be used for MS detection and chemical imaging, for inorganic and organic explosives as well as the speciation of the inorganic signatures of explosive devices. Analysis and proper spatial discrimination of HMX traces were demonstrated *via* imaging of fingerprints with this instrument. In-source CID functions to enhance collisions of ions with atmospheric gas, thereby reducing adducts and minimizing the interference. The whole system gives an opportunity for the isotopic and molecular speciation of inorganic components and presents further physicochemical information (Fig. **10**).

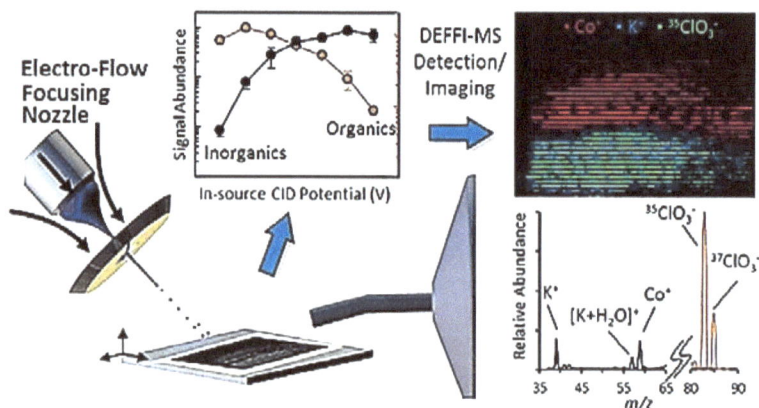

Fig. (10). Electroflow focusing and DEFFI-MS detection of explosives [Reprint from Ref. 138].

Colorimetric reactions and immunoassays are good for instant on-site analysis of liquid samples, SERS/IR presents a remote and non-destructive opportunity in identification and can be used for all phases, however, the best sensitivity can be obtained with GC/MS and LC-MS type techniques in liquid samples, DART-MS mostly in the gaseous phase, and techniques as DEFFI-MS in solid surfaces [139].

AI-MS is suitable for rapid, direct analysis of explosive residues on various surfaces, and no sample preparation is needed [36]. However, method sensitivity is influenced by the matrix composition and that the inter-day residual standard deviation may be up to 30%, a higher value than that of the existing validated methods. When these reasons are evaluated along with other non-technical factors, it can be concluded that AI-MS, at least for now, should be used only for the preselection of evidence rather than quantitative work to presents as a result in the courts. Since the probative analytical methods must pass the Daubert criteria [140] when AI-MS considered, the error rate should be quantifiable and acceptably low, and the error origins should be understood and controllable.

As an alternative to AI-MS, in case the quantitative result is needed, and extraction of sample is possible, flow injection analysis tandem mass spectrometry (FIA-MS/MS) can be used with its well known speed (<1 min/sample), and performance as an intermediate between AI-MS and conventional quantitative LC-MS/MS [36]. FIA is the method of analyzing samples by injecting them into the flowing solvent (mobile phase) [141]. In FIA–MS, which is termed loop injection, separation and identification occur within MS [142]. An analyte response is used in quantification.

To be able to analyze the target analytes in FIA-MS/MS, target analytes should be ionized, and the correct ion mode (positive or negative) should be used according to the chemical nature of the molecule. Sample preparation is very important in FIA, since the sample solvent has an impact on the ionization efficiency. If needed, an acid or base is used to obtain the desired pH or a dissolving assistant may be utilized to provide a complete dissolution. Since separation is not carried out for FIA, total ion chromatograms (TIC) can be used to monitor the produced ions and to confirm that the ions are from the target analytes. Since the determination with low concentrations are more difficult, it is recommended to work firstly with sample concentrations in sufficient ppm levels from the tens to maybe hundreds of ppm. A TIC for each ionization and a chromatogram for the mass number with the highest ion intensity is obtained from the analyses.

The newer generation of MS instruments gives a chance for the ambient techniques as FIA-MS/MS to get ever more capable [36]. Explosive detection methods with FIA APCI/ESI- MS/MS is suggested to be fast, sensitive, and selective and superior to most of the current methods (including those of EPA) in terms of sensitivity. Ostrinskaya *et al.* have developed a fast (1 min) method to determine 10 explosives from multiple explosive classes (organonitrates, organic peroxides, and oxidizer salts) urea nitrate, potassium chlorate, 2,4,6-trinitrotoluene, 2,4,6-trinitro phenylmethyl nitramine, triacetone triperoxide, hexamethylene triperoxide diamine, pentaerythritol tetranitrate, 1,3,5-trinitroperhydro-1,3,5-triazine, nitroglycerin, and octohydro-1,3,5,7-tetranit- o-1,3,5,7-tetrazocine in pg/mL levels using FIA-MS/MS and reported pg/mL levels sensitivities with 20-sec analysis times. The method is claimed to allow the integration of new analytes, since it is suitable to evolve for new challenges. Matrix effects (especially ion suppression) are also important for FIA because no chromatographic separation is carried out before ionization. According to Gosetti *et al.*, "suppression in APCI is due to the formation of the solid matter: a precipitate could form as the concentrations of both analyte and nonvolatile components increase with solvent evaporation." Also, matrix interferents may cause signal enhancement in APCI-MS/MS negative ionization mode and wrong assignments of the interfering peaks as the analyte may be possible with the triple-quadrupole mass spectrometer. In such cases, the use of the backup MS/MS transitions can be useful to decrease the uncertainties.

Ambient MS tools may also be combined with a hand-held IMS to be used in a wide range of common explosives [123]. Hyphenation of ambient techniques with miniature mass spectrometers provides fast, selective, and sensitive on-field explosives screening in airports, public buildings, *etc* [36].

The most crucial point in effective screening and confirmation is the detection of a wide scope of trace analytes among the potential interferences in different types of or extraordinary matrices, which are present in higher concentrations than the analytes. Today MS is used with automated systems without the requirement of highly skilled analysts. The improvements are directed towards the objective of detection with high selectivity/sensitivity, with systems that can be utilized in as many different matrices as possible and to analyze without extraction or with minimum and easy sample preparation in a short time. Moreover, miniaturizing advancing instruments to be utilized economic in-field detection is a standout amongst the most significant inspirations for the advancements around there.

On-site Detection

In the field or at the scene of a bombing, first-line tools such as colorimetric tests and portable instrumentation are used to make preliminary determinations that help investigators determine what explosives were used [31]. On site detection techniques vary according to the type of the sample; as well as vapors emanating from bulk solid explosives or trace particulates [143]. The vapor cloud from an explosive may contain molecular constituents, impurities, or degradation products from 1000 ppm to ppt levels. In such cases, detection using gas-phase molecular spectroscopies is possible. Molecular spectroscopic techniques can detect and identify explosive molecules in the vapor plume, but it will not be useful for many explosives with low vapor pressure. The molecules of most explosives (except nitroglycerin, EGDN-based dynamites, and TATP) are released in such a low amount into the gas phase that the detection is possible only at the close surface of the sample or the material.

Chemically specific spectroscopic probes urge direct sampling, near-surface measurements, and perhaps preconcentration. There is a characteristic issue with progressively increasing sensitive determination. False positives or false alarms may arise in ultrasensitive detection methods, because of the traces from the use of recreational firearms or medical application of nitroglycerin in heart diseases. Ultrasensitive detection methods would be tricky in military areas which are rich in explosives and could mislead in the analysis because of the chemical interferences.

UV-*Vis*, infrared, and microwave absorption are some of the techniques used for the detection of explosive molecules in the vapor phase. Fluorescence has also been used, however, specificity is low since UV-*Vis* and fluorescence bands in electronic spectra are broad.

GC interfaced with electron capture detectors (ECD) is used in direct sampling ionization. GC-MS is regarded as a standard on-site chemical analyzer even for complex mixtures; but bulky, expensive, and delicate equipment is needed. The improvement of microelectromechanical systems provide advances in developing fieldable mass spectrometers for a better adaptation in on-site analysis.

The main strategy presently utilized is IMS, which utilizes an electron source at ambient pressure to produce negative ions of explosive vapors and describe them by their drift times in a fixed electric field. GC-MS is more selective than IMS. However, IMS has been broadly conveyed for many reasons as being economical and easy instrumentation.

Particulate explosive material can be found around nonhermetically sealed explosives related with the contamination on the skin, explosive container, or nearby objects. The amount of particulate material is usually much greater than that of gaseous vapor. Trace analysis techniques as MS, IMS, electron capture, thermo-redox, and field ion spectrometry, surface acoustic wave sensors can identify molecules from small particles. Such techniques are usually hyphenated with front-end GC to prefractionate the molecules in incoming samples, so that the selectivity is enhanced. There are techniques among these that have been designed to act as "sniffers", for remote detection. Standoff detection is also possible with GC-MS and portable Fourier transform infrared (FTIR) using a laser beam and a telescope, however, decision should be taken regarding the background interference and the physical state of the explosive sample (*e.g.*, FTIR requires a solid sample) before selection of the technique to be used.

When a terrorist deals with an explosive, his or her hands and clothing will probably be contaminated with that explosive [1]. Thus, explosive traces from the hands of suspects has been studied many times in the literature. Explosive residues, may also be transferred to clothes and to other items, such as luggage, computers *etc.* Numerous trace explosive detectors have been produced for screening luggage and other objects at airports since the detection of trace amounts of explosives at airports is very important in combat with terrorism. Unfortunately, various advances that can be utilized for screening baggage can't be utilized for screening travelers for health reasons. Innovations are being created for mass transit systems where checkpoints are impractical, and while developing these techniques the limitations of mass transit should also be considered. Fast, dependable field detection of explosives is a top priority in both private and government security markets [32]. New studies in remote optical detection are increasing the identification power of governments and military organizations for explosives, biological agents and other threats [23]. Standoff detection is possible at distances from several centimeters up to a kilometer. Some standoff methods focus on chemical identification without making contact with the target or by probing the environment from a distance. Regularly, standoff detection alludes to the utilization of imaging techniques such as video, x-ray and identification software to identify suspicious packages, wire, fragmentation materials, or other evidences of IEDs or threats. The greatest difficulties in standoff discovery are increasing the distance where an effective identification can occur from, improving signal detection within environmental noise and interferences and screening of multiple mobile dangers.

The terrorist assault of September 11, 2001, and the attempted shoe bombing of American Airlines Flight 63 in December 2001, prompted reconsideration of the issues identified with airline security. Still, once more the increased concentration focused on the screening of luggage and travelers utilizing close-distance explosives detection. By the way, these safety precautions did not target the scenarios in which people show up in an open domain and a security choice must be made at a distance from the suspect. For such situations, standoff explosives detection is required, where physical separation protects people from possible damage. Of course, there are also disadvantages of such techniques. The trouble in standoff detection includes dynamic interference on the analyte signal, the potential for high false positives, and the need to determine a risk rapidly for taking a precaution.

On-site detecting of explosives may be dangerous because of the potential toxicity and the danger of a possible explosion [26]. So, any rapid and effective analysis technology which is capable of working at a safe distance will be extremely important in this application field.

To obtain a beneficial identification method for field detection of explosive devices; the technique should have high sensitivity, with the capacity to distinguish trace quantities of material, should be able to operate from a safe distance, should reach high specificity for explosives, in the presence of common interferences (sometimes termed 'chemical clutter') A good standoff technique should detect a weak signal in a noisy environment. This background noise is also often dynamic, thus, poor performance can be encountered in the field, in contrary to good results in controlled laboratory conditions. The speed of detection is critical when a potential threat is quickly drawing closer. Law implementation authorities of various countries around the globe need to promote the production of innovative detection systems to manage the issue of hidden explosives at airports, squares, railways, bus stations, *etc* [2].

In Pittcon 2019 Conference [26] it is stated that: *"The researches are needed in plume and aerosol dynamics; x-ray, microwave, infrared, and terahertz imaging and spectroscopy; neutron, gamma-ray, magnetic resonance, and magnetic-field systems; optical absorption and fluorescence; light detection and ranging (LIDAR), differential-absorption LIDAR (DIAL), and differential-reflectance LIDAR (DIRL); biosensors and biomimetic sensors; and microelectromechanical systems (MEMS). Researchers in chemical, mechanical, nuclear, and electrical engineering, bioengineering, chemistry, spectroscopy, applied physics, and optics should be involved in these efforts."*

Zhang *et al.* techniques [2] compared 4 spectroscopic detection techniques, as shown in Table **2**. As seen from the table, detection techniques with THz spectroscopy and IMS show superior performances in sensitivity, selectivity, and signal acquisition rate. But Raman has superior molecular selectivity to IMS.

Table 2 A short comparison of some spectroscopic detection techniques [Reprint from Ref 2].

Table 2. A short comparison of some spectroscopic detection techniques [2].

Methods	Sample Type	Sensitivity	Molecular Selectivity	Acquisition Time	*In Situ* Monitoring
THz spectroscopy	condensed	medium	good	fast	yes
LIBS	condensed	low	medium	fast	yes

(Table 2) cont.....

Methods	Sample Type	Sensitivity	Molecular Selectivity	Acquisition Time	*In Situ* Monitoring
Raman spectroscopy	condensed	medium	good	slow	yes
IMS	gas	high	medium	medium	yes

Terahertz technology in explosive detection from different sources has attracted much attention of the researchers with new attempts to develop better methods which have been partially successful [2]. The chance to use only a narrow terahertz window (only up to 3–6 THz) for spectrum collection is one of the reasons. Since several compounds may seem similar throughout this scale, this spectrum scale is not adequate to discriminate the different explosives and their matrix each.A broader bandwidth is expected to identify the characteristic features in the obtained spectrum of each molecule. Since the concentrations of explosives is generally in trace levels, and the influence of the surface quality is high, the success in acquiring accurate results requires a more sensitive arrangement in analysis and a cautious signal processing. To overcome the disadvantage of peak broadening, a waveguide-based time-domain spectroscopy (TDS) technology may be utilized where the crystal plane is highly directed to THz radiation.

LIBS is a technique which can be regarded as good in molecular selectivity, for a more reliable determination of explosives, however it has still has some disadvantages to overcome. In the explosive detection, interference from atmospheric oxygen and nitrogen is a classical problem, and a technology configured to prioritize residues can aid in overcoming the interference from the environment. The dual pulse LIBS may decrease the interference of the ambient atmosphere.

Further research through the use of portable Raman microscopes will also be useful so that testing whether the explosive particles can be carried out in the battlefield from the explosive mixture is a crucial aspect of the military actions. As a matter of fact, a microscope is not integrated into the current portable Raman spectrometers, so these can only be used in the detection of massive unexploded explosives.

Ion mobility spectrometry (IMS) is a widely used method for detecting explosives before detonation. It has the advantages of firmness, portability, and on-site use. However, studies performed in laboratory has definite limitations as low resolution, interference problem and limited selectivity, so they are not in every case straightforwardly applicable to the comprehensive screening of different explosives. For IMS, future innovations are directed to miniaturization, the substitution of non-radioactive ionization sources, and the improvement of its performance. Combining different strategies is a viable way to guarantee accurate detection. Hyphenation of different spectral techniques (as Raman, LIBS, *etc.*) in a single instrument can be applied sharing lazers, spectrometer, and optical paths. Furthermore, miniaturization will help to reduce the size of the instrument.

A number of explosive detection methods developed exist in the literature for humanitarian demining, groundwater and soil remediation, security screening, tactical forensic and intelligence, trace-evidence chemical imaging, and other forensic analyses [36].

Among the new technologies, The Transportation Security Administration began testing an innovative instrument using millimeter waves instead of harmful radiation — to scan metallic and non-metallic objects on a person's body [144]. The instrument is build to determine explosive suicide vests at Penn Station. Two types of machines were developed by Transportation Security Administration for emitting waves on people and screens from a distance without having them walk through a scanner and slowing them down. If there are any hidden items under clothing that could possibly be an improvised explosive device (IED) it will trigger an alarm on the operators' laptop. This technique was used in the security of some events as 2014 Super Bowl however since the development studies have not been completed, it is not yet under use at any transit areas.

Construction of portable CE instruments is also investigated, and a portable CE instrument with a coupled contactless conductivity detection C4D was implemented for the analysis of explosives, by Kobrin *et al.* [59]. CE C4D is a commonly-used detection technique for inorganic and organic species or biomolecules [145]. The electrodes are placed outside and wrap around the capillary tubes with a few mm detection gap, and form a capacitor with the electrolyte solution. Alternating current (ac) is applied to the first electrode, while the ac-current induced in the second electrode is measured, depending on the electric conductivity of the solution between the two electrodes. The detector cell is inexpensive and commercially available. However, the current challenge in advancing this technique concerns sampling and sample pretreatment. C4D is claimed to have superiority over the other instruments, from the point of being an economical technique with a relatively simple electronic circuity, and low power requirement. So it is regarded as very suitable for integration to portable CE devices.

The requirement for in situ MS analysis has increased over the years [146]. There are briefcase-sized MS systems [32], but the cost is an important factor in getting widespread use. Smaller instruments have simpler quadrupoles or ion traps.

GC-MS with electron ionization, one of the most preferred portable MS systems for on-site determination of explosives [32]. In GC-MS, at least 4 identification points are recommended to identify an analyte accurately [108]. One of them is the retention time. The other three are the minimum 3 characteristic fragment ions. Different techniques (like EI and CI) may be brought together, for example, the 2 ions obtained from EI method and then 2 with a CI method on the same instrument/column gives a total of 5 identification points. GC-MS became a common tool for monitoring organic pollutants in the environment [107]. The main reasons for preferring GC-MS in various fields are the characteristic distinctive property of enhanced molecular ion, reliability in identification, significantly increased spectrum of molecules with thermally lability and low-Volatility, rapid analysis, high sensitivity especially for the analytes that are difficult in analysis, *etc*. Its cost has decreased whereas the accuracy has significantly increased. Explosive detection systems are now an important component of all US airports. To improve the performance in homeland security and public health, conventional GC-MS units with the transmission quadrupole MS, as well as those with cylindrical ion trap (CIT-MS) and toroidal ion trap (T-ITMS) instruments have been modified for portability and near on-site detection of chemical warfare agents such as sarin, soman, and VX [14, 15].

Suitcase-type GC/MS systems with SPME fibers for sample introduction and low thermal mass GC assemblies for chromatographic separation exists. There is also a portable GC/MS product where the sample can be introduced directly through syringe injection, direct air sampling, purge-and-trap and headspace methods, and SPME [147] Miniaturized systems have been manufactured which use ESI and APCI for analysis of semi- and nonvolatile analytes. A compact quadrupole MS system targeting analysis of fume hoods can be coupled to HPLC or UPLC through both ESI and APCI sources, also for explosive analysis.

MS portable systems are expected to focus on not only reducing the size and weight but also maintaining the analytical performance while providing the possibility of performing MS analysis more simply. To produce portable mass spectrometers, beam mass analyzers such as the sector, time-of-flight (TOF), and quadrupole, along with ion trap, were miniaturized [146]. Smaller vacuum pumps, diaphragms and miniaturized power supplies also made it possible to develop portable MS systems.

Miniaturization decreases RP in TOF analyzers since the RP the flight length decreases. However, the use of a short flight tube that requires a low vacuum became possible with the use of energy-focusing devices, advances in high-speed electronics, and laser technology. The high resolution could be achieved through increasing the flight path, using spiral orbit trajectory in TOF analyzers or electrostatic multi-pass mirrors circulating the ions for a longer period in multipass TOF mass spectrometers.

Quadrupole mass filters require low vacuum and have small size and low weight and cost in comparison with other beam-type analyzers. Apart from their reduced size, an rf-only quadrupole can be used to guide ions through the fragmentation region and enable tandem MS experiments which bring more selectivity and sensitivity in comparison with instruments of single-stage fragmentation. Using microelectromechanical systems (MEMS), micro engineered quadrupole filters have been developed with a battery-powered portable mass spectrometer weighing 14.9 kg.[1] In addition, a miniaturized quadrupole mass spectrometer was used for the development of the first portable HPLC-MS system [148].

Ion traps are preferred in portable mass spectrometers because of its modest vacuum requirements (since it can operate at pressures higher than 10^{-3} Torr, where small pumping systems can be used) and intrinsic MS/MS capabilities [146, 148 - 150]. Also MS^n experiments in a single device is possible because the frequency is the base for mass analysis in ion traps. Toroidal rf ion trap mass spectrometer in the literature, presented higher capacity of ion storage, higher sensitivity and RP, compared to traditional 3D ion traps, and rectilinear ion trap (RIT) based mass analyzers has been shown to have a higher signal-to-noise ratio, resulting in higher trapping capacity and resolution more than 1000. Portable Mini 10 is an RIT based mass spectrometer weighing 10 kg and the Mini 11 weighing only 4 kg, which was developed with the capability to perform MS^n experiments operating under low power conditions.

Since the chromatography takes time and brings complexity to the procedure, it is known that fast ambient techniques including PSI (Paper Spray Ionization), direct analysis in real-time (DART) *etc.* has started to be preferred in the laboratory analysis [32]. In DART, ions produced from the sample (gas, liquid, or solid) are used directly for screening or sometimes confirmation of explosives in either unexploded or exploded form. DART analysis lasts a few seconds, and produces more information on explosive residues from many materials as luggage, plastics, metals, packaging, and paper currency. Airports are important places where explosives are screened continuously, using ion mobility instrumentation. Samples are usually collected through swabbing. If a positive alarm is obtained with instant screening with non-MS-based threat detection devices such as NIR, Raman, and IMS, further investigation as physical examination is performed. The suspected cases can be rapidly confirmed with a DART-MS [128, 129]. The ionization is typically performed as protonation for MS analysis. No solvent or sample extraction is needed, and using nitrogen as the ionizing gas offer convenience as a mobile laboratory equipment [32]. However, to our knowledge, DART has not been miniaturized or portable yet. Besides, it wouldn't be surprising that in the near future, portable DART-MS instruments hyphenated with techniques that would strengthen the on-field detection of explosives are adapted to the field analyses. DART provides a gentler ionization method as in LC-MS ionization, so it offers a chance to detect the analytes in lower concentrations than traditional GC-MS. Traditional GC-MS analyses may confirm the identity of explosives found in two locations, however DART-MS has superiority of providing an instant information on-site, including the matrix components and/or impurities, while gathering the evidences in crime scene investigation. As a disadvantage, since it does not have a chromatographic component, spectra may seem more complex for some sample types.

There are miniature instruments with DESI, however, because of high gas and solvent flows where the limited pumping capacity of small vacuum pumps may not be adequate, its use is limited in miniature mass spectrometers [151]. As a solution the ambient gas flow rate and the diameter of the inlet capillary may be decreased, although it can drop the signal strength. Or as a more effective solution, ions can be discontinuous.

Several portable MS systems that use ambient ionization methods were developed by the research groups of R. Graham Cooks and Zheng Ouyang at Purdue University [152]. Among these, a commercial, rugged MS system, the Flir Systems Atmospheric Inlet MS (AIMS), was also introduced to the market [147]. Cooks and coworkers and Flir Systems have been used desorption electrospray ionization (DESI) as a well-established method, hyphenated with portable MS. This hyphenation provides attractive superiorities, especially in trace surface analysis. High sensitivity and selectivity are accomplished due to the MS/MS capability. LODs of these systems range mostly from low- to sub-nanogram levels. Although not every ambient MS method can be coupled to portable instrumentation, several have been demonstrated or hold high promise for implementation. Cook *et al.* have incorporated the simplicity of PSI to portable MS for field investigation without the need for tiring or any sample preparation. Dr. Graham Cooks' and Dr Zheng Ouyang's groups from Purdue University (West Lafayette, USA) reported several applications on the miniature PSI-MS instruments which they have developed. With this versatile hyphenation, data can be acquired immediately. The advantages of both paper chromatography and electrospray ionization is assembled on a a triangularly-cut paper or porous substrate as a disposable ionization source [153]. Different sampling strategies may be used: deposition via pipetting, swabbing or dipping, and the swab itself can be used as a surface-transfer medium. Ionization type is similar to that of ESI: Sample, solvent and high voltage are applied to the paper, ions are formed in solution and the strong electric field at the tip of the paper results in the formation of a plume of charged droplets which are desolvated and directed to the mass spectrometer. PSI is a more commonly used ionization technique with miniature systems [151]. Ionization could be achieved at low voltages (~3V) with the integration of carbon nanotubes into the paper, forming a certain amount of charged droplets from the uncharged droplets. Since no pneumatics or extensive sample preparation is necessary, it is one of the simplest and preferrable techniques which can be hyphenated to portable MS systems [153]. PSI is widespread in the forensic community, also in the fast detection of explosives and chemical warfare agents. Cartridges have been manufactured for PSI, to hold paper, and to increase the reproducibility and robustness of the technique [151]. Paper spray cartridges for analysis focusing on certain points on the Mini 12, 3D-printed fast wetting cartridges and continuous supply of solvent, and cartridges for SPE integration were produced. Routine subsequent analyses is possible with a serial array of paper triangles on a moving stage, and even droplet analysis became possible with the integration of paper spray into a gravity-driven microfluidic chip. In related trials, the chemical structure of the paper was modified to gain a hydrophobic nature for reduced wicking and extended analyte signal or instead of paper, a membrane ("membrane electrospray"), plant material ("leaf splash"), aluminum foil, or a funnel shaped metal or paper channel was used.

Splashes on various materials as sharp metal tips, therapeutic swabs, medicinal infusion needles, wooden tips, C18-covered cutting edges,or a glass microscope slide were analyzed. It was shown that when the paper which was used to collect a sample was coated with carbon nanotubes, the voltage required was diminished 1000 times, the signal was sharper, and the instrument could catch much more sensitive molecules. The dramatic loss in power required methods a decrease in battery size and experimental cost. The nanotubes work like minor antennas that make a strong electric field from an extremely little voltage. One volt over a couple of nanometers makes an electric field proportionate to 10 million volts over a centimeter. The diminished voltage provides a gentler method than the standard PaperSpray™ ionization procedures. The voltage used for ambient ionization through spraying on a carbon nanotube (CNT) impregnated paper surface is as low as ≥3V [154]. In such a condition, organic molecules give simple high quality mass spectra without fragmentation in the positive or negative ion modes unlike the conventional field ionization. Fragile molecules and complexes are able to hold together in this gentle ionization medium. CNTs coated on the paper surface are responsible for high electric fields The analytical performance was displayed on some volatile and non-volatile analytes and a series of matrices

One of the most important problems during MS miniaturization is the requirement of a high vacuum environment for high resolution, but since high sensitivity and low-power-consumption is expected from these instruments [150], the design of vacuum and atmospheric pressure interface (API) is one of the key challenges, which largely determine the size and power consumption of an MS system, as well as its analytical performances. Miniaturization has been reported, including for turbo-molecular, membrane, rotary, cryogenic and ion-getter pumps to provide adequate vacuum for these instruments [146]. To transfer the ionized analytes to the high vacuum environment of the mass analyzer, a discontinuous atmospheric pressure interface (DAPI) that opens periodically was used, and the ions were sent to MS in a pulsed way, with small flows. These instruments also have adopted different atmospheric pressure interfaces (APIs), as membrane inlets (MIs), and continuous atmospheric pressure interface (CAPI). MI based systems have limitations in sensitivity for nonvolatile samples [150]. DAPI can be coupled with ambient ionization sources for different analytes. However, generally, it has low stability and analysis speed with a duty cycle of only 1%, determined by the pulsed sampling introduction using a mechanical valve. CAPI with a miniature differential pumping system enables continuous sampling introduction into miniature mass spectrometers, provides adequate performance, high stability and rapid analysis speed for complex samples using ambient ionization sources.

A probability-based source-to-source comparison, similar to how DNA is used for individualization is suggested for predicting the origin of the explosives. Such an approach requires not only the ability to discriminate two samples from different sources, but also an evaluation of whether samples did, or did not, originate from the same source [32]. For this method, an understanding of background variation and the proper selection of control samples are needed. To collect and analyze samples of known origins is necessary in order to obtain a probabilistic value, such as a likelihood ratio, to evaluate the explosive evidences. The raw materials and fillers used to produce explosives often originate from many locations. It is also suggested that; "A process-based model that can relate the isotope ratios of an explosive to its precursors could further increase the reliability in the estimation of the origin of explosives.

RESULTS

Recently, explosive screening and confirmation is a trendy field of analytical studies, because of the increased need for homeland security against terrorism and warfare, as well as their environmental and toxicological aspect [36]. To detect explosives in post-blast debris is important from the point of finding the type of the bomb and the source and to reach the suspects [78]. Explosives are used frequently today in military and terroristic applications. They also pose a permanent and increasing concern on human and ecosystem health. These chemicals are highly toxic, some carcinogenic, and their detection in rural areas, surrounding military bases and weapon training facilities, became imperious.

Post-detonation explosive compounds are notoriously difficult to detect and analyze. The vapor pressure of most explosives is very low [35]. For example, TNT vapor pressure is about 5.93×10^{-6} mbar at room temperature, resulting in a maximum concentration of approximately 6 ppb in air. Therefore, analytical methods that rely on the injection of air spaces either need to use large volumes of samples or may have very high limits of detection (LODs). Techniques based on raising the temperature to achieve higher vapor pressures often lead to thermal degradation of materials, thus inhibiting their detection. Because of this, in MS techniques relying on GC separation, decomposition of some molecules may occur because of the thermal lability of the explosives, and the detection spectrum cannot expand much generally and at this point, LC-MS and LC-MS/MS offer distinct advantages over GC-MS and GC-MS/MS for laboratory analysis. LC-MS/MS is also superior to LC-MS in selectivity and sensitivity in complex matrices and multi-analyte determination in one run. The negative detection error using the MS-MS technique is partially eliminated [30]. To detect trace concentrations of explosives before their decomposition, rapid, sensitive, less laborious and economical techniques are needed. LC–MS/MS is well-suited for such analyses regarding relatively polar molecules and heat labile compounds [155], which may degrade at the high temperatures typically used in GC [156]. LC-MS/MS methods, including simultaneous analysis of RDX, HMX and PETN or RDX, HMX and TNT in soil, or a wide spectrum explosives in the soil are also limited in the literature. Some existing methods in soil are, directed to qualitative purposes, tedious and time-consuming, have longer times of chromatographic analysis, require higher amounts of solvent/sample ratios and some of them lack recovery or some lack validation. So studies on the development of new fast methods on the multi-analyte determination with high analytical performance will be very useful especially in finding out the type of explosive used in the post-blast debris in a terroristic attack. More effective ways of environmental-friendly extraction techniques with minimum solvent use, determining a larger spectrum of explosives, chromatographic separation hyphenated with MS/UV techniques, are required to detect, confirm and quantify these trace level explosive compounds at ppb and sub-ppb levels [78]. Worldwide, there is a demand to lower the analytical detection limit to nano or pico level while keeping a high efficiency, good reproducibility, low cost and rapidity of the chosen method. There are many aspects of an analytical work, where improvements can be achieved as; (a) sample preparation, (b) compounds separation and detection and (c) ionization techniques (paper spray ionization [157, 158], alternating current corona discharge APCI [159], dual ionization sources ESI/APCI [158], dielectric barrier discharge [160], *etc.*). The instrumental aspects, advantages and disadvantages are compared in Table **3**.

Recently, new research is recently performed for the development of an interface that can connect IMS or other tandem MS strategies to the orbitrap mass analyzers [161]. So that structural identification studies for explosives would be performed better using ion mobility and MS resolution combination. A modified reverse entry ion source (REIS) was produced to generate a cold ion source interfaced *via* an RF-only octupole ion guide to the higher-energy collisional dissociation (HCD) cell of the orbitrap. The full performance was retained with this configuration and was tested on the analysis of ubiquitin, the trimeric transmembrane ammonia transport channel (AmtB) complex, as well as the products of lipid binding to AmtB. In the instrument, ions were introduced into a heated capillary using nanoflow ESI (n-ESI), then focused using an RF ion funnel. Ions exiting the funnel pass through a skimmer region before being focused through a ring electrode and finally enter an RF-only octupole ion guide. The ions entering *via* REIS are indiscriminately guided and trapped in the HCD cell. The HCD cell retains its functionality: ions can be stored and transmitted to the C-Trap as well as being heated or fragmented before mass analysis.

Even this design may keep most of the functions of the orbitrap, the capability of utilizing automatic gain control (AGC) to limit space charge effects in the orbitrap disappeared *via the* interface. This innovation gives the opportunity to produce future instrument modifications where additional gas-phase coupled techniques such as IMS can be adapted to the existing instrument, bringing a chance to use, in explosive analysis in complicated matrices.

High resolution Time-of-Flight (ToF) and triple quadrupole technology present a unique ease-of-use, accuracy, and sensitivity. Triple quadrupole use in detection provides two-dimensional separations, and thus offers an analysis option in less than 1 minute with no need for column separation [36]. Reliable results for courts can be produced with FIA APCI/ ESI MS/MS, performing proper validation with internal standard and external standard calibration curves.

A stand-off device should have multiple sensors of various types, or should provide characteristic ion channels in MS, which increase the analyte spectrum that can be caught in the environment. If there are indications of explosives that each sensor type can find among the environmental interferences, these sensors may be added also to increase selectivity/specificity and sensitivity.

Since distinguishing an analyte in a noisy environment remains a challenge, studies should target baseline determination in ambient conditions and variations in detecting at ambient conditions in real time.

Table 3. Comparison of disadvantages and advantages of the MS techniques used in explosive analysis in the literature.

Table 3. Comparison of disadvantages and advantages of the MS techniques used in explosive analysis in the literature.

Technique		Advantages	Limitations
GC-MS TD-GC-MS LD-GC-MS		suitable for volatile and thermally labile molecules combined with sorptive sampling techniques are often used as an alternative to sensitivity limitations of liquid extraction techniques high robustness suitable for automation	Problems with molecules of high volatility and thermally labile. Adsorption of polar explosives in the chromatographic system, causing a decrease in response and peak tailing manual performing of stir-bar removing from the sample, rinsing and drying. SBSE desorption step is disadvantageous compared to LC, because of the complexity in the automation. the obligation of low injection volume as 1-2uL
GC-MS/MS		suitable for volatile and thermally labile molecules superior selectivity than GC-MS methods (it can differentiate the analytes from interferences and co-eluting compounds) higher sensitivity than GC-MS	
GC-GC-TOF/MS With disposable PDMS twister sorptive sample		a wider range of trace analytes in a complex matrix during a single run better resolution (increased selectivity) higher sensitivity compared to the traditional GC PDMS loop is cheaper than SBSE faster and easier sample introduction cryofocusing is not necessary narrow peaks for volatile and semivolatile analytes even with the splitless mode in laborious suitable for thermally labile molecules different explosive compounds (RDX, PETN, and TNT) can be detected as well as taggant molecules and plasticizers the detection of less volatile compounds may also be possible with the low elution temperature low LOD fast analysis	
LVI-GC-MS		superior for semi-volatile and thermally labile compounds higher sensitivity through introducing an efficient volume as; tens of microliters or to even hundreds of microliters of sample	reaching large volumes of solvent vapor from a sample to the MS system
PTV-LVI-GC-MS		the smaller internal volume of PTV inlet in splitless mode and the need for low temperature decreases thermal decomposition sample vapors will remain shorter in the hot vaporizing chamber improved detection limits	Analyte protectants are recommended to decrease the adsorption of polar analytes, peak tailing and to receive higher responses, Problems can be experienced with thermolabile analytes
Portable GC-MS		enhanced molecular ions reliability in identification significantly increased analyte range with thermally lability and low-volatility rapid analysis high sensitivity	Chromatography takes time and bring complexity to the procedures, DART has started to be preferred.
PTV-LVI-doublecolumnGC-MS(using a flame detector to remove solvent overload with the second column)		Very high sensitivities (as 0,1 pg/mL) LOD in soil: 0.5ng/g with a linear dynamic range of 2 orders magnitude The continuous separation of the solvent from the compounds through the entire chromatographic column prevented the loss of volatile analytes along with the solvent	
LC-MS		superiority over GC-MS and GC-MS/MS in the analysis of relatively polar molecules and heat-labile compounds It can be used with different extraction techniques as SBSE, UAE, SPE, *etc.*	
LC-MS/MS		provide superiority over GC-MS and GC-MS/MS in the analysis of relatively polar molecules and heat-labile compounds superior to LC-MS in selectivity and sensitivity in complex matrices and multi-analyte determination in one run. False negatives are partially eliminated Very low LODs with triple quadrupole MS. It can be used with different extraction techniques as SBSE, UAE, SPE, *etc.*	high costs lengthy sample analysis needs well-qualified staff explosive analysis methods developed are limited in the literature sensitivity decreases in ESI mode for nitroaromatics.
LC-APCI-MS/MS		less variation of matrix effects than ESI less fragmentation higher signals the adducts of nitramines, nitrate esters and nitroaromatic compounds are detected with APCI(-)	
LC-ESI-MS/MS		ESI (-) is better in detecting nitro esters, nitramines, especially for nitrate, chloride, formate and acetate adducts.	Matrix effects are higher
LC-QToF-MS		better identification and information of analytes and accurate mass even at sub-ppm levels, because of high RP better selectivity via enabling the fragmentation of pre-selected ions and the identification of compounds using their product ion spectra identification possibility of unknown compounds even there are no available reference standards	
LC/ NI-APPI-ToF/MS		highly sensitive for the analysis of DNB, DNT, TNB, TNT, HMX and RDX good reproducibility high precision	LOQ values were two orders of magnitude lower compared to APCI-LC/MS
HPLC-(PDA)-LTQ Orbitrap FTMS		acquisition of full scan MSn (n-stage fragmentations, n = 1 − n) spectra with an LTQ detector at high speed and/or with the Orbitrap having ultra-high mass resolution and high mass accuracy LC-Orbitrap-MS gives the chance to identify unknown compounds even there are no available reference standards	LOD is lower than triple quadrupole MS
CE-MS		Simple instrumental setup great versatility economic requires solvent and sample in lower amounts high on-field analysis potential	only the molecules which are charged or can be transformed to charged forms can be analyzed

(Table 3) cont.....

Technique		Advantages	Limitations
UV-VUV-MS		the isomer selectivity of vacuum UV absorption spectroscopy without destroying the analytes more powerful characterization of volatile organic analytes in complex matrices maybe tried for the volatile organic compounds can be combined with the above mentioned PTV-LVI-GC technique above, in order to enlarge the scale of detectable compounds and to provide a more precise identification.	
IMS	Remote detection allow real-time monitoring widely used for the fast, sensitive on-field detection of trace levels of nitroaromatic explosives at airports, public buildings, *etc.*		lower selectivity limited linear range low resolution high concentrations of explosives and complex matrices may lead to saturation and contamination of the instrument, which causes interferences in quantitative determinations
IMS coupled with MS		have higher sensitivity detects a wide range of explosives	
AMBIENT MS	Real time analysis directly from the surface (solids, liquids, gases) Much wider range of applications much higher selectivity and sensitivity minimal sample preparation the possibility of chemical imaging APLD and AP-LIAD evaporate explosives and detect them directly in vapour phase		Chemical imaging, for inorganic and organic explosives as well as the speciation of the inorganic signatures of explosive devices Used only for the preselection of evidence rather than quantitative work to presents as a result in the courts
MALDI-TOF-MS		Compared to chromatography-based MS instruments: Simple tolerant of sample impurities fast acquisition needs small sample volumes samples are prepared more easily and can be automated easily for routine analyses special ionization properties and tracking all ions in the sample in the same spectrum may allow the detection of degradation products.	For a good ionization, the choice of the matrix is crucial. It can detect low g/mL levels. Since no chromatographic separation exists, only mass-based identification may lead to misinterpretation.
DART-MS		solvent-free DART can also be combined with a second heating component, combining thermal desorption and ionization for better control. A fast transient heating would desorb volatile compounds at low temperatures and inorganic salts via elevating the temperature subsequently. DART analysis lasts a few seconds (shorter than 15-30 min GC elution) and produces more information on explosive residues from many materials as luggage, plastics, metals, packaging, and paper currency It offers chance to detect the analytes in lower concentrations than traditional GC-MS. DART-MS has superiority of providing instant information on-site, including the matrix components and/or impurities, during the crime scene investigation	difficulties for the analysis of low volatility inorganic analytes, which require elevated temperatures for efficient desorption directed to identification rather than quantitation
JHTD-DART-MS		Presented an efficient desorption of organic and inorganic explosives at their respective optimal temperature for each, followed by rapid cooling, maintaining high throughput. detection of inorganic compounds (as ammonium nitrate, calcium ammonium nitrate, potassium chlorate, potassium perchlorate, and potassium nitrite, *etc.*) even at lower nanogram levels Thermal desorption of elemental inorganics are claimed to be possible through exchanging the wire material	
IRTD-DART-MS		Detection of both inorganic and organic explosives from wipe collected samples detects volatile, semivolatile organic explosives as well as non-volatile oxidizer based explosives. Detects organic and inorganic oxidizer based inorganic compounds at ng and sub-ng levels.	Further improvement in sensitivity is limited by the thermal properties of the commercial wipes.
DART-Orbitrap MS combined with Raman		providing spectral fingerprints that are orthogonal in nature pg to ng levels fast screening comprehensive characterization differentiation of explosive particulate samples	
DESI-MS		More number of analytes can be analyzed than that of DART. DESI-MS can analyze ambient vapors at 3 m distance, where PETN and TNT can be determined in quite low amounts	interferences which can be caused by the suppression of ionization product of the matrix effects in the samples adding an additional step in the preparation of the sample is often required.
PSI-MS	.	fast detection from surfaces combines extraction and ionization in one step, using solvent-wetted paper cut in triangle matrices as; wooden toothpicks swabs or the sample itself, such as in leaf spray, can be used instead of papers. present new chances for security screening	ionization suppression in the analysis of complex samples low sensitivity at trace levels low spray stability narrow dynamic range

(Table 3) cont.....

Technique		Advantages	Limitations
DCBI-MS, DBDI-MS, SESI-MS, SIFT-MS, PTR-MS	.	No sample preparation Detection limits at usually ppt or lower levels use of multiple reagent ions decrease the interferences of SIFT-MS and increase the specificity in real-time gas analysis. RP of PTR-TOF-MS enables identification of isobaric compounds in complex mixtures a dual SESI and DBDI source was used with MS analyzer for real-time fast vapor analysis and used for canine training in the detection of explosives.	switchable reagent ion (SRI-PTR-MS) can be used. But it doesn t have the instantaneous reagent ion switching capabilities, stability, specificity, and soft ionization of SIFT-MS
DEFFI-MS		chemical imaging, for inorganic and organic explosives speciation of the inorganic signatures of explosive devices	
FIA-MS/MS		in case the quantitative result is needed, and extraction of sample is possible, it can be used with its well known speed and performance as an intermediate between AI-MS and LC-MS/MS	sample preparation is very important, since the sample solvent has an impact on the ionization efficiency. determination with low concentrations are more difficult
FIA-APCI/ESI-MS/MS		fast selective superior to most of the current methods (including those of EPA) in terms of sensitivity (pg/mL) Suitable to evolve for the integration of new analytes	matrix effects are important because no chromatographic separation exists wrong assignments of the interfering peaks as the analyte may be possible with the triple-quadrupole mass spectrometer.
LC-NMR/TOF-MS	NMR and MS data compared and correlated when using identical sample preparations allows improved opportunities for catching new unknown explosives offers new avenues for the routine quantification of explosive derivatives It is least-destructive (prepared sample can be used in repetitive NMR experiments or other analytical techniques followed by NMR).		studies are directed to detection of environmental contaminants stability issues in their MS analyses or weak or overlapping signals may cause a limitation for some cases

The studies are going on for presenting faster methods to the literature, with wider spectrum of energetic molecules in more matrices. Production of new, innovative detection strategies and the improvement of existing techniques are of vital importance [1]. Miniaturization, portability, low cost, fast analysis, operation simplicity, high selectivity/sensitivity are among the expected improvements. The future studies under the light of the information obtained from the recent experiences are planned to be carried out in various matrices with the higher number of explosives in order to catch the molecules that exist in the post-blast debris in one run.

THE CURRENT NEEDS & FUTURE PROSPECTS

New analytical methods with improved performances should be developed for faster and more sensitive determination of explosives. The studies are going on about explosive detection techniques with fast, reliable, and precise real-time analysis of the trace levels of explosives with high resolution without sample preparation. Recently, the techniques based on spectroscopic determination are the most advanced strategies for explosive analysis. Thus, it seems that they will most probably also play a crucial role also in the future.

The new developments in the synthesis of new and some powerful explosives some of which leave little or no trace behind is also another challenge in explosive analysis techniques: Besides the various nitro compounds utilized as military and business explosives, fuels, and propellants [162] on which the analysis methods are mostly focused, there is a continuous search for more powerful and less shock-sensitive derivatives. Persistent endeavors have been made to grow new energetic materials having good thermal stability, impact and shock insensitivity, better performance, economical and green syntheses, to meet the future requirements of military and space applications [163]. High explosives have a potent exothermic reactivity, which is attractive for both military and business applications [164].

Molecular perovskites have attracted growing attention from the point of design and production of new and more powerful energetic materials [165]. They mimic the structure of inorganic perovskites but have at least one molecular component. Zhang, who is one of the founders of octanitrocubane, said, "The perovskite structure is favorable for the stability. The new explosive materials have rather high thermal stability and low impact sensitivity, which is good for their storage and transportation." Besides the classical explanation that energy release is due to the breaking and recombination of chemical bonds during the denotation reaction, the release of the structural tension in the frameworks also make a great contribution to the explosive performances. Zhang indicated:"Different from traditional design focusing on the intramolecular functional groups, we emphasize the inter-molecular assembly in the specified crystal." said he admitted: "Low-cost fuel and oxidizer are integrated into highly-symmetric ternary crystals.". Zhang and his co-worker Chen emphasized the flexibility of such a design and added these words: "Molecular components with suitable shape and size can assemble into such kinds of ternary crystals to optimize the oxygen balance, crystal density and so on. Better explosive performances are expected". Recently, the attention is focused on high-density organic explosives, which contain all of the elements required for combustion to gaseous products in the absence of air. Energetic materials with the strained rings and cages as nitrocubanes are a promising new class of explosives [162, 166]. From these strained cages, especially octanitrocubane (ONC) and heptanitrocubane are the most powerful non-nuclear high-energetics currently known [167]. ONC is a cubane derivative with stressed rings and cages, with a high nitrogen content and high density between 1.98 - 2.06 g/cm^3, varying according to the nitro group orientation. Since it is shock insensitive, it provides a safer handling. Since ONC has no hydrogen, no H_2O forms when it burns; it leaves little or no visible smoke behind the rocket and these "low-signature" rockets will be difficult to track [162]. Its synthesis and characterization was performed by Eaten *et al.* in 2002, however, there is no information if a batch-synthesis has been made recently, whether it is used already by the military, as well as no trade of it seems to be carried out by any manufacturer. If batch synthesis will start to be made and used in terroristic attacks, because of the tracking or the analysis in post-blast debris will be more difficult than other explosive residues, this will be a serious challenge in the detection for analytical and forensic chemists.

There are analysis studies with the NMR technique in the literature. If powerful techniques as NMR are hyphenated with MS techniques, it will be possible to catch many explosive types through characterization of the molecules tracing the signals of common functional groups found in organic explosives, which may escape from the method that the analyst develops. NMR technique is used with hyphenation with MS techniques recently in the literature. For example, Gowda *et al.* introduced a new method where the metabolites in a serum sample are quantified using a recently developed NMR method [168] and then used as a reference for absolute metabolite quantitation using MS. This NMR-guided-MS quantitation approach is simple and easy to implement and offers new avenues for the routine quantification of explosive derivatives using MS. The demonstration that NMR and MS data can be compared and correlated when using identical sample preparations allows improved opportunities to exploit their combined strengths for catching new unknown explosives through the identification of the functional groups and molecules. However, one should keep in consideration that stability issues in their MS analyses or weak or overlapping signals may cause a limitation in some cases.

NMR is advantageous to combine with MS. It is least-destructive (the prepared sample can be used in repetitive NMR experiments, or the sample can be analyzed by other analytical techniques followed by the analysis in NMR). There are studies in the literature using NMR technique hyphenated with MS. However, these are directed to the detection of environmental contaminants [169]. In an environmental study where HPLC-SPE-NMR/TOF-MS (high-performance liquid chromatography coupled to solid-phase extraction and nuclear magnetic resonance and time-of-flight mass spectrometry) was used in the characterization of xenobiotic contaminants in groundwater. The structural elucidation can be achieved through the use of 1H NMR spectra of postcolumn SPE enriched compounds, together with accurate mass measurements. 2,4,6-TNT, RDX, HMX, degradation products of TNT (1,3,5-TNB), (2-A-4,6-DNT), 3,5-DNP, 3,5-DNA, 2,6-dinitroanthranile, and 2-hydroxy-4,6-dinitrobenzonitrile were detected in low ngmL^{-1} concentrations.

New technological and methodological improvements, along with the formation of explosive databases and automated algorithms for data processing/analysis, will greatly increase the qualitative and quantitative strength of the explosive analysis. Hyphenated techniques and methodologies that provide structural and/or elemental information about an explosive is strongly recommended.

Explosive screening and confirmation remain an active and wide-spectrum area of research in recent years, due to increased demand for homeland security against terrorist and warfare threats, as well as environmental monitoring of these molecules. The most important point in an effective determination system of explosive residues is the simultaneous detectability of a broad range of compounds at trace levels among the possible interferences coming from the various complex and sometimes unique matrices, which exist in much higher concentrations than the explosive analytes. Today MS is being transformed into automated systems that do not require high skilled staff, which can be used in robust matrices and environments. The developments are progressing to the target of obtaining high selectivity/sensitivity for the analytes, to be able to use the technique in as many different matrices as possible and to be able to analyze without or with a concise and simple sample preparation methods. Speed, ease is important in forensic analysis. Furthermore, miniaturization for evolving instruments to be used in-field analysis with low cost is one of the most important motivations for the developments in this area. As a final note, regarding all the issues related to explosive analysis in the laboratory; it is obvious that the world doesn't need much more waste and it is giving an alarm. Thus, all steps of an analytical determination method, especially in forensics and environmental analysis, should be converted to green techniques avoiding expensive and disposable materials as much as possible.

LIST OF ABBREVIATIONS

AD-XRD	angle dispersive X-ray diffraction
AGC	Automatic gain control
AI-MS	Ambient Ionization-Mass Spectrometry
2AmDNT	2-amino-4,6-dinitrotoluene
AN	ammonium nitrate
ANFO	(AN/FO = nitrate/fuel oil)
APCI	atmospheric pressure chemical ionization
API	atmospheric pressure ion sources
APLD	Ambient pressure laser desorption
AP-LIAD	ambient pressure laser-induced acoustic desorption
Au TNPs	gold triangular nanoprisms
CE	capillary electrophoresis
CIT	cylindrical ion trap

CL-20	Hexanitrohexaazaisowurtzitane
CNTs	carbon nanotubes
COTS	commercial off-the-shelf
CPE	cloud-point extraction
DART/MS	Direct Analysis in Real Time
DBDI	dielectric barrier discharge ionization
DCBI	corona beam ionisation source
DEFFI	electro-flow focusing ionization
DIAL	differential-absorption
DIRL	differential-reflectance
DMNB	2,3-dimethyl-2,3-dinitrobutane
DNB	Dinitrobenzene
2,4 DNT 2	2,4-dinitrotoluene 6-DNT:2,6-dinitrotoluene
DTIMS	Drift tube ion mobility coupled to MS
ECD	Electron Capture Detector
EC	ethylcentralite
ED- XRD	X-Ray Diffraction Spectrometry
EGDN	Ethylene glycol dinitrate
ESI	electrospray ionization
ESIS	Externally sampled internal standard
FIA-MS/MS	flow injection analysis tandem mass spectrometry
FTIR	Fourier transform infrared
HRMS	High resolution mass spectrometric
FNA	fast neutron analysis
FTMS	Fourier Transform Mass Spectrometer
GC-ECD	Gas chromatography electron capture detector
GC-NCI	Gas chromatography negative chemical ionization
GC-NPD	Gas chromatography nitrogen phosphorus detector
HCD	higher-energy collisional dissociation
HMEs	homemade explosives
HMTD	hexamethylene triperoxide diamine
HMX	octahydro-1,3,5,7-tetranitro-1,3,5,7-tetrazocine
HNS	hexanitrostilbene
IED	improvised explosive device
ICR	have ultra-high RP.

IMS	Ion mobility spectrometry
IRTD	Infrared thermal desorption
JHTD	Joule heating thermal desorption
IT	ion trap
LIBS	Lazer Induced Breakdown Spectroscopy
LIDAR	light detection and ranging
LIT	linear ion trap
LLE	liquid-liquid extraction
LSPR	localized surface plasmon resonance
LTQ	Linear Trap Quadrupole
MEKP	Metil etil keton peroks
MEMS	microelectromechanical systems
MSPE	magnetic solid-phase extraction
MTBE	Methyl tert-butyl ether
NB	Nitrobenzene
NG	nitroglycerine
NI-APPI-MS	negative ion atmospheric pressure photoionization mass spectrometry
NT	Nitrotoluene
ONC	octanitrocubane
PDA	photodiode array
PETN	Pentaerythritol tetranitrate
RDX	1,3,5-trinitroperhydro-1,3,5-triazine
TDS	conventional thermal desorption system
TEA	Thermal Energy Analyser
TOF	Accu-Time of Flight
SBSE	stir bar sorptive extraction
SDME	single drop microextraction
SERS	Surface Enhanced Raman Spectroscopy
SFC	supercritical fluid chromatography
SFE	supercritical fluid extraction
SLE	solid liquid extraction
SPE	solid phase extraction
SPME	solid phase microextraction
Q	quadrupole
QuEChERS	Quick Easy Cheap Effective Rugged Safe

REIS	reverse entry ion source
RF	Radio Frequency
RP	Resolution Power
PCI	positive chemical ionization
PDMS	polydimethylsiloxane
PSPME	planar solid phase microextraction
PVP-DVB	polyvinyl-pyrrolidone-divinylbenzene
PTV-LVI	programmed temperature vaporizing -large volume injection
PTR	proton transfer reaction
PSI	paper spray ionization
SESI	electrospray ionization
SIFT	selected-ion-flow-tube
TATP	triacetone triperoxide
TIC	total ion chromatograms
T-ITMS	toroidal ion trap
TNA	thermal neutron analysis
TNB	Trinitrobenzene
TNT	2,4,6-trinitrotoluene
THz	TerahertzTHz
TDS	THz spectroscopy is Time Domain Spectroscopy
ToF	High resolution Time-of-Flight
UAE	Ultrasonic assisted extraction
VUV	vacuum ultraviolet

CONSENT FOR PUBLICATION

Not applicable.

CONFLICT OF INTEREST

The authors confirm that this chapter contents have no conflict of interest.

ACKNOWLEDGEMENTS

Declared none.

REFERENCES

[1]   Lees, H. Development of Analysis Methods to Detect the Use of Explosives and Chemical Warfare Agents, Tallinn University of Technology Doctoral Thesis 45/2018, 2018.

[2]   Zhang, W.; Tang, Y.; Shi, A.; Bao, L.; Shen, Y.; Shen, R.; Ye, Y. Recent developments in spectroscopic techniques for the detection of explosives. *Materials (Basel),* **2018**, *11*(8), 1364.
[http://dx.doi.org/10.3390/ma11081364] [PMID: 30082670]

[3]   Brust, G.M.H. *Chemical profiling of explosives, Analytical Chemistry including its applications in Forensic Science (HIMS, FNWI)*; CPI Wöhrmann: Amsterdam, **2014**.

[4]   Eiceman, G.A.; Schmidt, H. *Advances in Ion Mobility Spectrometry of Explosives.Aspects of Explosives Detection*; Elsevier B. V.: Amsterdam, **2009**, p. 171.
[http://dx.doi.org/10.1016/B978-0-12-374533-0.00009-X]

[5]   Üzer, A.; Can, Z.; Akın, I.; Erçağ, E.; Apak, R. 4-Aminothiophenol functionalized gold nanoparticle-based colorimetric sensor for the determination of nitramine energetic materials. *Anal. Chem.,* **2014**, *86*(1), 351-356.
[http://dx.doi.org/10.1021/ac4032725] [PMID: 24299426]

[6]   Üzer, A.; Sağlam, S.; Tekdemir, Y.; Ustamehmetoğlu, B.; Sezer, E.; Erçağ, E.; Apak, R. Determination of nitroaromatic and nitramine type energetic materials in synthetic and real mixtures by cyclic voltammetry. *Talanta,* **2013**, *115*, 768-778.
[http://dx.doi.org/10.1016/j.talanta.2013.06.047] [PMID: 24054661]

[7]   Swider, J.R. Optimizing accu time-of-flight/direct analysis in real time for explosive residue analysis. *J. Forensic Sci.,* **2013**, *58*(6), 1601-1606.
[http://dx.doi.org/10.1111/1556-4029.12276] [PMID: 24117693]

[8]   Emmons, E.D.; Tripathi, A.; Guicheteau, J.A.; Fountain, A.W., III; Christesen, S.D. Ultraviolet resonance Raman spectroscopy of explosives in solution and the solid state. *J. Phys. Chem. A,* **2013**, *117*(20), 4158-4166.
[http://dx.doi.org/10.1021/jp402585u] [PMID: 23656503]

[9]   Yinon, J. Trace analysis of explosives in water by gas chromatography—mass spectrometry with a temperature-programmed injector. *J. Chromatogr. A,* **1996**, *742*(1-2), 205-209.
[http://dx.doi.org/10.1016/0021-9673(96)00261-0]

[10]   Šesták, J.; Večeřa, Z.; Kahle, V. novel portable device for fast analysis of energetic materials in the environment. *Proceeding of New Trends in Research of Energetic Materials NTREM,* **2012**, *2012*(Part II), 860-865.

[11]   Saha, S; Mandal, M K; Chen, L C; Ninomiya, S; Shida, Y; Hiraoka, K. *Trace level detection of explosives in solution using Leidenfrost phenomenon assisted thermal desorption ambient mass spectrometry,* **2013**, *2*(Spec Iss), S0008.
[http://dx.doi.org/10.5702/massspectrometry.S0008]

[12]   Xu, X.; Koeberg, M.; Kuijpers, C.J.; Kok, E. Development and validation of highly selective screening and confirmatory methods for the qualitative forensic analysis of organic explosive compounds with high performance liquid chromatography coupled with (photodiode array and) LTQ ion trap/Orbitrap mass spectrometric detections (HPLC-(PDA)-LTQOrbitrap). *Sci. Justice,* **2014**, *54*(1), 3-21.
[http://dx.doi.org/10.1016/j.scijus.2013.08.003] [PMID: 24438773]

[13]   Holmgren, E.; Carlsson, H.; Goede, P.; Crescenzi, C. Determination and characterization of organic explosives using porous graphitic carbon and liquid chromatography-atmospheric pressure chemical ionization mass spectrometry. *J. Chromatogr. A,* **2005**, *1099*(1-2), 127-135.
[http://dx.doi.org/10.1016/j.chroma.2005.08.088] [PMID: 16213509]

[14]   Xu, X.; van de Craats, A.M.; de Bruyn, P.C. Highly sensitive screening method for nitroaromatic, nitramine and nitrate ester explosives by high performance liquid chromatography-atmospheric pressure ionization-mass spectrometry (HPLC-API-MS) in forensic applications. *J. Forensic Sci.,* **2004**, *49*(6), 1171-1180.
[http://dx.doi.org/10.1520/JFS2003279] [PMID: 15568687]

[15]   Jiang, G. *Simultaneous UPLC/MS Analyses of Explosive Compounds, Application Note 51829*;

Thermo Fisher Scientific: Texas, **2010**.

[16]   Hewitt, A.D.; Jenkins, T.F.; Walsh, M.E.; Walsh, M.R.; Taylor, S. RDX and TNT residues from live-fire and blow-in-place detonations. *Chemosphere,* **2005**, *61*(6), 888-894.
       [http://dx.doi.org/10.1016/j.chemosphere.2005.04.058] [PMID: 15964048]

[17]   Marple, R.L.; LaCourse, W.R. Application of photoassisted electrochemical detection to explosive-containing environmental samples. *Anal. Chem.,* **2005**, *77*(20), 6709-6714.
       [http://dx.doi.org/10.1021/ac050886v] [PMID: 16223260]

[18]   Marple, R.L.; Lacourse, W.R. A platform for on-site environmental analysis of explosives using high performance liquid chromatography with UV absorbance and photo-assisted electrochemical detection. *Talanta,* **2005**, *66*(3), 581-590.
       [http://dx.doi.org/10.1016/j.talanta.2004.11.034] [PMID: 18970024]

[19]   Sánchez, C.; Carlsson, H.; Colmsjö, A.; Crescenzi, C.; Batlle, R. Determination of nitroaromatic compounds in air samples at femtogram level using C18 membrane sampling and on-line extraction with LC-MS. *Anal. Chem.,* **2003**, *75*(17), 4639-4645.
       [http://dx.doi.org/10.1021/ac034278w] [PMID: 14632075]

[20]   Furton, K.G.; Almirall, J.R.; Bi, M.; Wang, J.; Wu, L. Application of solid-phase microextraction to the recovery of explosives and ignitable liquid residues from forensic specimens. *J. Chromatogr. A,* **2000**, *885*(1-2), 419-432.
       [http://dx.doi.org/10.1016/S0021-9673(00)00368-X] [PMID: 10941688]

[21]   Halasz, A.; Groom, C.; Zhou, E.; Paquet, L.; Beaulieu, C.; Deschamps, S.; Corriveau, A.; Thiboutot, S.; Ampleman, G.; Dubois, C.; Hawari, J. Detection of explosives and their degradation products in soil environments. *J. Chromatogr. A,* **2002**, *963*(1-2), 411-418.
       [http://dx.doi.org/10.1016/S0021-9673(02)00553-8] [PMID: 12187997]

[22]   Batlle, R.; Carlsson, H.; Holmgren, E.; Colmsjö, A.; Crescenzi, C. On-line coupling of supercritical fluid extraction with high-performance liquid chromatography for the determination of explosives in vapour phases. *J. Chromatogr. A,* **2002**, *963*(1-2), 73-82.
       [http://dx.doi.org/10.1016/S0021-9673(02)00136-X] [PMID: 12188003]

[23]   Coffey,   M.   Chemical   and   Explosives   Detection,   American   Physical   Society.
       2019.https://www.aps.org/about/governance/task-force/counter-terrorism/coffey.cfm

[24]   Suna, S. European Pressphoto Agency, Photo number: 5315728. **2016**.http://www.epa.eu

[25]   Brown, T.L.; LeMay, H.E., Jr; Bursten, B.E. *Chemistry: The Central Science*; Pearson Australia Group Pty. Ltd.: Frenchs Forest, NSW, **2014**.

[26]   Materials, I.E.; Residues, A.P-E. Identifying Explosive Materials and Analyzing Post-Explosion Residues – The Rise in Handheld Devices Pittcon Conference & Expo (March 5-9, 2017, McCormick Place, Chicago, IL. USA March 17 - 21, 2017, Pennsylvania Convention Center, Philadelphia, Pennsylvania, USA.

[27]   Naik, V; Patil, K C High Energy Materials A Brief History and Chemistry of Fireworks and Rocketry Resonance. (5)*20*, 431-444.**2015**,

[28]   Explosives Technical Background and Glossary of Ordnance Terms From Picatinny Arsenal, Historic American Engineering Record **1984**. (HAER) No. NJG36,

[29]   López-López, M.; García-Ruiz, C. Infrared and Raman spectroscopy techniques applied to identification of explosives. *Trends Analyt. Chem.,* **2014**, *54*, 36-44.
       [http://dx.doi.org/10.1016/j.trac.2013.10.011]

[30]   Şen, N.; Üzek, U.; Aksoy, Ç.; Bora, T.; Atakol, O. Farklı Yapıdaki Organik Patlayıcı Maddelerin LC-MS-MS ile Belirlenmesi, SDU Journal of Science (E-Journal), 2015, 10 (1): 95-106. DergiPark.
       https://dergipark.org.tr/tr/pub/sdufeffd/issue/11281/134823/

[31]   Corbin, I. *Analysis of Improvised Explosives by Electrospray Ionization - Mass Spectrometry and*

Microfluidic   Techniques.   FIU   Electronic   Theses   and   Dissertations. **2016**.https://digitalcommons.fiu.edu/etd/2551

[32]   DePalma, A. Lab Manager, *Emerging Technologies for Detecting, Identifying, and Analyzing Hazardous   Materials,*   Laboratory   Hazards   and   Risks. **2017**.https://www.labmanager.com/insights/2017/06/emerging-technologies-f- r-detecting-identifyi-g-and-analyzing-hazardous-materials#.Xh2y_8gzaM8

[33]   Robinson, E.L.; Sisco, E.; Staymates, M.E.; Lawrence, J.A. A new wipe-sampling instrument for measuring the collection efficiency of trace explosives residues. *Anal. Methods,* **2018**, *10*(2), 204-213. [http://dx.doi.org/10.1039/C7AY02694C] [PMID: 29881468]

[34]   Kirchner, M.; Matisová, E.; Hrouzková, S.; Húšková, R. Fast GC and GC-MS analysis of explosives. *Petrol. Coal,* **2007**, *49*(2), 72-79.

[35]   Badjagbo, K.; Sauvé, S. Mass Spectrometry for Trace Analysis of Explosives in Water. *Crit. Rev. Anal. Chem.,* **2012**, *42*(3), 257-271. [http://dx.doi.org/10.1080/10408347.2012.680332] [PMID: 22746321]

[36]   Ostrinskaya, A.; Kunz, R.R.; Clark, M.; Kingsborough, R.P.; Ong, T.H.; Deneault, S. Rapid Quantitative Analysis of Multiple Explosive Compound Classes on a Single Instrument *via* Flow-Injection Analysis Tandem Mass Spectrometry. *J. Forensic Sci.,* **2019**, *64*(1), 223-230. [http://dx.doi.org/10.1111/1556-4029.13827] [PMID: 29797696]

[37]   Davies, A.G.; Burnett, A.D.; Fan, W.; Linfield, J.E.H.; Cunningham, E. Terahertz spectroscopy of explosives and drugs. *Mater. Today,* **2008**, *11*, 18-26. [http://dx.doi.org/10.1016/S1369-7021(08)70016-6]

[38]   Zapata, F.; de la Ossa, M.Á.F.; Gilchrist, E.; Barron, L.; García-Ruiz, C. Progressing the analysis of Improvised Explosive Devices: Comparative study for trace detection of explosive residues in handprints by Raman spectroscopy and liquid chromatography. *Talanta,* **2016**, *161*, 219-227. [http://dx.doi.org/10.1016/j.talanta.2016.05.057] [PMID: 27769399]

[39]   Zapata, F.; García-Ruiz, C. Determination of Nanogram Microparticles from Explosives after Real Open-Air Explosions by Confocal Raman Microscopy. *Anal. Chem.,* **2016**, *88*(13), 6726-6733. [http://dx.doi.org/10.1021/acs.analchem.6b00927] [PMID: 27281604]

[40]   Zapata, F.; García-Ruiz, C. Analysis of different materials subjected to open-air explosions in search of explosive traces by Raman microscopy. *Forensic Sci. Int.,* **2017**, *275*, 57-64. [http://dx.doi.org/10.1016/j.forsciint.2017.02.032] [PMID: 28324768]

[41]   Ewing, R.G.; Atkinson, D.A.; Eiceman, G.A.; Ewing, G.J. A critical review of ion mobility spectrometry for the detection of explosives and explosive related compounds. *Talanta,* **2001**, *54*(3), 515-529. [http://dx.doi.org/10.1016/S0039-9140(00)00565-8] [PMID: 18968275]

[42]   Harvey, S.D.; Ewing, R.G.; Waltman, M.J. Selective sampling with direct ion mobility spectrometric detection for explosives analysis. *Int. J. Ion Mobil. Spectrom.,* **2009**, *12*, 115-121. [http://dx.doi.org/10.1007/s12127-009-0031-z]

[43]   Choi, S-S.; Son, C.E. Analytical method for the estimation of transfer and detection efficiencies of solid state explosives using ion mobility spectrometry and smear matrix. *Anal. Methods,* **2017**, *9*, 2505-2510. [http://dx.doi.org/10.1039/C7AY00529F]

[44]   Zhao, M.; Yu, H.; He, Y. A dynamic multichannel colorimetric sensor array for highly effective discrimination of ten explosives. *Sens. Actuators B Chem.,* **2019**, *283*, 329-333. [http://dx.doi.org/10.1016/j.snb.2018.12.061]

[45]   Wu, Z.; Duan, H.; Li, Z.; Guo, J.; Zhong, F.; Cao, Y.; Jia, D. Multichannel Discriminative Detection of Explosive Vapors with an Array of Nanofibrous Membranes Loaded with Quantum Dots. *Sensors (Basel),* **2017**, *17*(11), 2676.

[http://dx.doi.org/10.3390/s17112676] [PMID: 29156627]

[46]   Benson, S.J.; Lennard, C.J.; Maynard, P.; Hill, D.M.; Andrew, A.S.; Roux, C. Forensic analysis of explosives using isotope ratio mass spectrometry (IRMS)--discrimination of ammonium nitrate sources. *Sci. Justice,* **2009**, *49*(2), 73-80.
[http://dx.doi.org/10.1016/j.scijus.2009.04.005] [PMID: 19606584]

[47]   Grimm, B.L.; Stern, L.A.; Lowe, A.J. Forensic utility of a nitrogen and oxygen isotope ratio time series of ammonium nitrate and its isolated ions. *Talanta,* **2018**, *178*, 94-101.
[http://dx.doi.org/10.1016/j.talanta.2017.08.105] [PMID: 29136920]

[48]   Brown, K.E.; Greenfield, M.T.; McGrane, S.D.; Moore, D.S. Advances in explosives analysis--part II: photon and neutron methods. *Anal. Bioanal. Chem.,* **2016**, *408*(1), 49-65.
[http://dx.doi.org/10.1007/s00216-015-9043-1] [PMID: 26446898]

[49]   Skoulakis, A.; Androulakis, G.C.; Clark, E.L. A portable pulsed neutron generator. *Int J Mod Phys: Conference Series,* **2014**, *27*, 1460127-1460128.

[50]   Brown, K.E.; Greenfield, M.T.; McGrane, S.D.; Moore, D.S. Advances in explosives analysis--part I: animal, chemical, ion, and mechanical methods. *Anal. Bioanal. Chem.,* **2016**, *408*(1), 35-47.
[http://dx.doi.org/10.1007/s00216-015-9040-4] [PMID: 26462922]

[51]   Liyanage, T.; Rael, A.; Shaffer, S.; Zaidi, S.; Goodpaster, J.V.; Sardar, R. Fabrication of a self-assembled and flexible SERS nanosensor for explosive detection at parts-per-quadrillion levels from fingerprints. *Analyst (Lond.),* **2018**, *143*(9), 2012-2022.
[http://dx.doi.org/10.1039/C8AN00008E] [PMID: 29431838]

[52]   Lin, H.; Suslick, K.S. A colorimetric sensor array for detection of triacetone triperoxide vapor. *J. Am. Chem. Soc.,* **2010**, *132*(44), 15519-15521.
[http://dx.doi.org/10.1021/ja107419t] [PMID: 20949933]

[53]   Berliner, A.; Lee, M-G.; Zhang, Y.; Park, S.H.; Martino, R.; Rhodes, P.A. Yi G-R, Lim SH. A patterned colorimetric sensor array for rapid detection of TNT at ppt level. *Rsc Adv,* **2014**, *4*(21), 10672-10675.
[http://dx.doi.org/10.1039/c3ra47152g]

[54]   Kostesha, N.; Alstrom, T.S.; Johnsen, C.; Nielsen, K.A.; Jeppesen, J.O.; Larsen, J.; Boisen, A.; Jakobsen, M.H. Multi-colorimetric sensor array for detection of explosives in gas and liquid phase. *Proc. SPIE,* **2011**, •••801880181H--> [https://doi.org/10.1117/12.883895].
[http://dx.doi.org/10.1117/12.883895]

[55]   Johns, C.; Shellie, R.A.; Potter, O.G.; O'Reilly, J.W.; Hutchinson, J.P.; Guijt, R.M.; Breadmore, M.C.; Hilder, E.F.; Dicinoski, G.W.; Haddad, P.R. Identification of homemade inorganic explosives by ion chromatographic analysis of post-blast residues. *J. Chromatogr. A,* **2008**, *1182*(2), 205-214.
[http://dx.doi.org/10.1016/j.chroma.2008.01.014] [PMID: 18221942]

[56]   McCord, B.R.; Hargadon, K.A.; Hall, K.E.; Burmeister, S.G. Forensic analysis of explosives using ion chromatographic methods. *Anal. Chim. Acta,* **1994**, *288*, 43-56.
[http://dx.doi.org/10.1016/0003-2670(94)85115-8]

[57]   Gaurav, D.; Malik, A.K.; Rai, P.K. High-performance liquid chromatographic methods for the analysis of explosives. *Crit. Rev. Anal. Chem.,* **2007**, *37*, 227-268.
[http://dx.doi.org/10.1080/10408340701244698]

[58]   Schramm, S.; Vailhen, D.; Bridoux, M.C. Use of experimental design in the investigation of stir bar sorptive extraction followed by ultra-high-performance liquid chromatography-tandem mass spectrometry for the analysis of explosives in water samples. *J. Chromatogr. A,* **2016**, *1433*, 24-33.
[http://dx.doi.org/10.1016/j.chroma.2016.01.011] [PMID: 26777783]

[59]   Kobrin, E.G.; Lees, H.; Fomitšenko, M.; Kubáň, P.; Kaljurand, M. Fingerprinting postblast explosive residues by portable capillary electrophoresis with contactless conductivity detection. *Electrophoresis,* **2014**, *35*(8), 1165-1172.

[http://dx.doi.org/10.1002/elps.201300380] [PMID: 24375169]

[60]   Hutchinson, J.P.; Evenhuis, C.J.; Johns, C.; Kazarian, A.A.; Breadmore, M.C.; Macka, M.; Hilder, E.F.; Guijt, R.M.; Dicinoski, G.W.; Haddad, P.R. Identification of inorganic improvised explosive devices by analysis of postblast residues using portable capillary electrophoresis instrumentation and indirect photometric detection with a light-emitting diode. *Anal. Chem.,* **2007**, *79*(18), 7005-7013.
[http://dx.doi.org/10.1021/ac0708792] [PMID: 17705451]

[61]   Hopper, K.G.; Leclair, H.; McCord, B.R. A novel method for analysis of explosives residue by simultaneous detection of anions and cations *via* capillary zone electrophoresis. *Talanta,* **2005**, *67*(2), 304-312.
[http://dx.doi.org/10.1016/j.talanta.2005.01.037] [PMID: 18970170]

[62]   de la Ossa, M.Á.; Torre, M.; García-Ruiz, C. Determination of nitrocellulose by capillary electrophoresis with laser-induced fluorescence detection. *Anal. Chim. Acta,* **2012**, *745*, 149-155.
[http://dx.doi.org/10.1016/j.aca.2012.07.032] [PMID: 22938620]

[63]   Fernández de la Ossa, M.Á.; Ortega-Ojeda, F.; García-Ruiz, C. Discrimination of non-explosive and explosive samples through nitrocellulose fingerprints obtained by capillary electrophoresis. *J. Chromatogr. A,* **2013**, *1302*, 197-204.
[http://dx.doi.org/10.1016/j.chroma.2013.06.034] [PMID: 23845757]

[64]   Bailey, C.G.; Wallenborg, S.R. Indirect laser-induced fluorescence detection of explosive compounds using capillary electrochromatography and micellar electrokinetic chromatography. *Electrophoresis,* **2000**, *21*(15), 3081-3087.
[http://dx.doi.org/10.1002/1522-2683(20000901)21:15<3081::AID-ELPS3081>3.0.CO;2-R] [PMID: 11001203]

[65]   Brensinger, K.; Rollman, C.; Copper, C.; Genzman, A.; Rine, J.; Lurie, I.; Moini, M. Novel CE-MS technique for detection of high explosives using perfluorooctanoic acid as a MEKC and mass spectrometric complexation reagent. *Forensic Sci. Int.,* **2016**, *258*, 74-79.
[http://dx.doi.org/10.1016/j.forsciint.2015.11.007] [PMID: 26666592]

[66]   Perr, J.M.; Furton, K.G.; Almirall, J.R. Gas chromatography positive chemical ionization and tandem mass spectrometry for the analysis of organic high explosives. *Talanta,* **2005**, *67*(2), 430-436.
[http://dx.doi.org/10.1016/j.talanta.2005.01.035] [PMID: 18970185]

[67]   Smith, M.; Collins, G.E.; Wang, J. Microscale solid-phase extraction system for explosives. *J. Chromatogr. A,* **2003**, *991*(2), 159-167.
[http://dx.doi.org/10.1016/S0021-9673(03)00234-6] [PMID: 12741596]

[68]   Li, X. Agilent Technologies, Application Note 5990-6322EN_5989_5672, 2010. Environmental Protection Agency, U.S., Method 3650B. , **2010**.

[69]   Anilanmert, B.; Aydin, M.; Apak, R.; Avci, G.Y.; Cengiz, S. A fast liquid chromatography tandem mass spectrometric analysis of PETN (pentaerythritol tetranitrate), RDX (3, 5-trinitro-1, 3, 5-triazacyclohexane) and HMX (octahydro-1, 3, 5, 7-tetranitro-1, 3, 5, 7-tetrazocine) in soil, utilizing a simple ultrasonic-assisted extraction with minimum solvent. *Anal. Sci.,* **2016**, *32*(6), 611-616.
[http://dx.doi.org/10.2116/analsci.32.611] [PMID: 27302580]

[70]   Avci, G.F.Y.; Anilanmert, B.; Cengiz, S. Rapid and simple analysis of trace levels of three explosives in soil by liquid chromatography—tandem mass spectrometry. *Acta Chromatogr.,* **2017**, *29*(1), 45-56.
[http://dx.doi.org/10.1556/1326.2017.29.1.03]

[71]   Şener, H.; Anilanmert, B.; Cengiz, S. A fast method for monitoring of organic explosives in soil: a gas temperature gradient approach in LC–APCI/MS/MS. *Chem. Pap.,* **2017**, *71*(5), 971-979.
[http://dx.doi.org/10.1007/s11696-016-0042-2]

[72]   Katilie, C.J.; Simon, A.G.; DeGreeff, L.E. Quantitative analysis of vaporous ammonia by online derivatization with gas chromatography - mass spectrometry with applications to ammonium nitrate-based explosives. *Talanta,* **2019**, *193*, 87-92.
[http://dx.doi.org/10.1016/j.talanta.2018.09.099] [PMID: 30368302]

[73]   Brown, H.; Kirkbride, K.P.; Pigou, P.E.; Walker, G.S. New developments in SPME Part 2: Analysis of ammonium nitrate-based explosives. *J. Forensic Sci.,* **2004**, *49*(2), 215-221.
[http://dx.doi.org/10.1520/JFS2003219] [PMID: 15027534]

[74]   Psillakis, E.; Kalogerakis, N. Solid-phase microextraction *versus* single-drop microextraction for the analysis of nitroaromatic explosives in water samples. *J. Chromatogr. A,* **2001**, *938*(1-2), 113-120.
[http://dx.doi.org/10.1016/S0021-9673(01)01417-0] [PMID: 11771829]

[75]   U.S. EPA. "Method 8330B (SW-846): Nitroaromatics, Nitramines, and Nitrate Esters by High Performance Liquid Chromatography (HPLC)," Revision 2. Washington, DC. **2006**.http://www.epa.gov/ solidwaste/hazard/testmethods/pdfs/8330b.pdf

[76]   Mu, R.; Shi, H.; Yuan, Y.; Karnjanapiboonwong, A.; Burken, J.G.; Ma, Y. Fast separation and quantification method for nitroguanidine and 2,4-dinitroanisole by ultrafast liquid chromatography-tandem mass spectrometry. *Anal. Chem.,* **2012**, *84*(7), 3427-3432.
[http://dx.doi.org/10.1021/ac300306p] [PMID: 22414071]

[77]   Sisco, E.; Najarro, M.; Bridge, C.; Aranda, R., IV Quantifying the degradation of TNT and RDX in a saline environment with and without UV-exposure. *Forensic Sci. Int.,* **2015**, *251*, 124-131.
[http://dx.doi.org/10.1016/j.forsciint.2015.04.002] [PMID: 25909992]

[78]   Veresmortean, C.; Covaci, A. Hyphenated and non-hyphenated chromatographic techniques for trace level explosives in water bodies–a review. *Int. J. Environ. Anal. Chem.,* **2018**, *98*(5), 387-412.
[http://dx.doi.org/10.1080/03067319.2018.1478969]

[79]   Cortada, C.; Vidal, L.; Canals, A. Determination of nitroaromatic explosives in water samples by direct ultrasound-assisted dispersive liquid-liquid microextraction followed by gas chromatography-mass spectrometry. *Talanta,* **2011**, *85*(5), 2546-2552.
[http://dx.doi.org/10.1016/j.talanta.2011.08.011] [PMID: 21962682]

[80]   DeTata, D.A.; Collins, P.A.; McKinley, A.J. A comparison of solvent extract cleanup procedures in the analysis of organic explosives. *J. Forensic Sci.,* **2013**, *58*(2), 500-507.-->
[https://doi.org/10.1111/1556-4029.12035].
[http://dx.doi.org/10.1111/1556-4029.12035] [PMID: 23278326]

[81]   Rapp-Wright, H.; McEneff, G.; Murphy, B.; Gamble, S.; Morgan, R.; Beardah, M.; Barron, L. Suspect screening and quantification of trace organic explosives in wastewater using solid phase extraction and liquid chromatography-high resolution accurate mass spectrometry. *J. Hazard. Mater.,* **2017**, *329*, 11-21.
[http://dx.doi.org/10.1016/j.jhazmat.2017.01.008] [PMID: 28119193]

[82]   Kabir, A.; Holness, H.; Furton, K.G.; Almirall, J.R. Recent advances in micro-sample preparation with forensic applications. *Trends Analyt. Chem.,* **2013**, *45*, 264-279.
[http://dx.doi.org/10.1016/j.trac.2012.11.013]

[83]   Guerra, P.; Lai, H.; Almirall, J.R. Analysis of the volatile chemical markers of explosives using novel solid phase microextraction coupled to ion mobility spectrometry. *J. Sep. Sci.,* **2008**, *31*(15), 2891-2898.
[http://dx.doi.org/10.1002/jssc.200800171] [PMID: 18666175]

[84]   Thomas, J.L.; Donnelly, C.C.; Lloyd, E.W.; Mothershead, R.F., II; Miller, M.L. Development and validation of a solid phase extraction sample cleanup procedure for the recovery of trace levels of nitro-organic explosives in soil. *Forensic Sci. Int.,* **2018**, *284*, 65-77.
[http://dx.doi.org/10.1016/j.forsciint.2017.12.018] [PMID: 29353810]

[85]   Berton, P.R.; Regmi, B.P.; Spivak, D.A.; Warner, I.M. Ionic liquid-based dispersive microextraction of nitrotoluenes in water samples. *Mikrochim. Acta,* **2014**, *181*, 1191-1198.
[http://dx.doi.org/10.1007/s00604-014-1261-2]

[86]   Babaee, S.; Beiraghi, A. Micellar extraction and high performance liquid chromatography-ultra violet determination of some explosives in water samples. *Anal. Chim. Acta,* **2010**, *662*(1), 9-13.

[http://dx.doi.org/10.1016/j.aca.2009.12.032] [PMID: 20152259]

[87] Sun, X.; Liu, Y.; Shaw, G.; Carrier, A.; Dey, S.; Zhao, J.; Lei, Y. Fundamental study of electrospun pyrene–polyethersulfone nanofibers using mixed solvents for sensitive and selective explosives detection in aqueous solution. *ACS Appl. Mater. Interfaces,* **2015**, *7*(24), 13189-13197.
[http://dx.doi.org/10.1021/acsami.5b03655] [PMID: 26030223]

[88] Padrón, M.E.; Afonso-Olivares, C.; Sosa-Ferrera, Z.; Santana-Rodríguez, J.J. Microextraction techniques coupled to liquid chromatography with mass spectrometry for the determination of organic micropollutants in environmental water samples. *Molecules,* **2014**, *19*(7), 10320-10349.
[http://dx.doi.org/10.3390/molecules190710320] [PMID: 25033059]

[89] Psillakis, E.; Naxakis, G.; Kalogerakis, N. Detection of TNT-contamination in spiked-soil samples using SPME and GC/MS. *Global Nest. Int. J.,* **2000**, *2*(3), 227-236.-->
[https://doi.org/10.30955/gnj.000153].

[90] Fayazi, M.M.; Ghanei-Motlagh, M.; Taher, M.A. Combination of carbon nanotube reinforced hollow fiber membrane microextraction with gas chromatography-mass spectrometry for extraction and determination of some nitroaromatic explosives in environmental water. *Anal. Methods,* **2013**, *5*(6), 1474-1480.
[http://dx.doi.org/10.1039/c3ay25644h]

[91] Reyes-Gallardo, E.M.; Lasarte-Aragonés, G.; Lucena, R.; Cárdenas, S.; Valcárcel, M. Hybridization of commercial polymeric microparticles and magnetic nanoparticles for the dispersive micro-solid phase extraction of nitroaromatic hydrocarbons from water. *J. Chromatogr. A,* **2013**, *1271*(1), 50-55.
[http://dx.doi.org/10.1016/j.chroma.2012.11.040] [PMID: 23237711]

[92] Khezeli, T.; Daneshfar, A. Development of dispersive micro-solid phase extraction based on micro and nano sorbents. TrAC. *Trends Analyt. Chem.,* **2017**, *89*, 99-118.-->
[https://doi.org/10.1016/j.trac.2017.01.004].
[http://dx.doi.org/10.1016/j.trac.2017.01.004]

[93] Wooding, M.; Rohwer, E.R.; Naudé, Y. Comparison of a disposable sorptive sampler with thermal desorption in a gas chromatographic inlet, or in a dedicated thermal desorber, to conventional stir bar sorptive extraction-thermal desorption for the determination of micropollutants in water. *Anal. Chim. Acta,* **2017**, *984*, 107-115.
[http://dx.doi.org/10.1016/j.aca.2017.06.030] [PMID: 28843553]

[94] Ghani, M.; Ghoreishi, S.M.; Azamati, M. In-situ growth of zeolitic imidazole framework-67 on nanoporous anodized aluminum bar as stir-bar sorptive extraction sorbent for determining caffeine. *J. Chromatogr. A,* **2018**, *1577*, 15-23.
[http://dx.doi.org/10.1016/j.chroma.2018.09.049] [PMID: 30316613]

[95] Baltussen, E.; Cramers, C.A.; Sandra, P.J. Sorptive sample preparation -- a review. *Anal. Bioanal. Chem.,* **2002**, *373*(1-2), 3-22.
[http://dx.doi.org/10.1007/s00216-002-1266-2] [PMID: 12012168]

[96] Matisová, E.; Hrouzková, S. Analysis of endocrine disrupting pesticides by capillary GC with mass spectrometric detection. *Int. J. Environ. Res. Public Health,* **2012**, *9*(9), 3166-3196.
[http://dx.doi.org/10.3390/ijerph9093166] [PMID: 23202677]

[97] Bader, N. *Stir bar sorptive extraction as a sample preparation technique for chromatographic analysis: An overview. Asian J. Nanosci. Mater.,* **2018**.1(2. pp. 52-103), 56-62.

[98] Franc, C.; David, F.; de Revel, G. Multi-residue off-flavour profiling in wine using stir bar sorptive extraction-thermal desorption-gas chromatography-mass spectrometry. *J. Chromatogr. A,* **2009**, *1216*(15), 3318-3327.
[http://dx.doi.org/10.1016/j.chroma.2009.01.103] [PMID: 19233369]

[99] Bicchi, C.; Iori, C.; Rubiolo, P.; Sandra, P. Headspace sorptive extraction (HSSE), stir bar sorptive extraction (SBSE), and solid phase microextraction (SPME) applied to the analysis of roasted Arabica coffee and coffee brew. *J. Agric. Food Chem.,* **2002**, *50*(3), 449-459.

[http://dx.doi.org/10.1021/jf010877x] [PMID: 11804511]

[100] Bicchi, C.; Cordero, C.; Rubiolo, P.; Sandra, P. Stir bar sorptive extraction (SBSE) in sample preparation from heterogeneous matrices: determination of pesticide residues in pear pulp at ppb (ng/g) level. *Eur. Food Res. Technol.*, **2003**, *216*(5), 449-456.
[http://dx.doi.org/10.1007/s00217-003-0669-4]

[101] Tienpont, B.; David, F.; Bicchi, C.; Sandra, P. High capacity headspace sorptive extraction. *J. Microcolumn Sep.*, **2000**, *12*(11), 577-584.
[http://dx.doi.org/10.1002/1520-667X(2000)12:11<577::AID-MCS30>3.0.CO;2-Q]

[102] Dean, J.R. *Extraction Techniques in Analytical Sciences*; John Wiley & Sons: West Sussex, **2010**, Vol. 34, .

[103] Gómez, M.J.; Herrera, S.; Solé, D.; García-Calvo, E.; Fernández-Alba, A.R. Automatic searching and evaluation of priority and emerging contaminants in wastewater and river water by stir bar sorptive extraction followed by comprehensive two-dimensional gas chromatography-time-of-flight mass spectrometry. *Anal. Chem.*, **2011**, *83*(7), 2638-2647.
[http://dx.doi.org/10.1021/ac102909g] [PMID: 21388147]

[104] Tranchida, P.Q.; Franchina, F.A.; Dugo, P.; Mondello, L. Comprehensive two-dimensional gas chromatography-mass spectrometry: Recent evolution and current trends. *Mass Spectrom. Rev.*, **2016**, *35*(4), 524-534.
[http://dx.doi.org/10.1002/mas.21443] [PMID: 25269651]

[105] Shi, X.; Wang, S.; Yang, Q.; Lu, X.; Xu, G. Comprehensive two-dimensional chromatography for analyzing complex samples: recent new advances. *Anal. Methods*, **2014**, *6*(18), 7112-7123.
[http://dx.doi.org/10.1039/C4AY01055H]

[106] Barron, L.; Gilchrist, E. Ion chromatography-mass spectrometry: a review of recent technologies and applications in forensic and environmental explosives analysis. *Anal. Chim. Acta*, **2014**, *806*, 27-54.
[http://dx.doi.org/10.1016/j.aca.2013.10.047] [PMID: 24331039]

[107] Chauhan, A.; Goyal, M.K.; Chauhan, P. GC-MS technique and its analytical applications in science and technology. *J. Anal. Bioanal. Tech.*, **2014**, *5*(6), 222.
[http://dx.doi.org/10.4172/2155-9872.1000222]

[108] ATF-LS-E09 Detection of Explosives by Gas Chromatography/Mass Spectrometry (GC/MS), Authority: Technical Leader, March. **2018**.https://www.atf.gov/file/127846/download

[109] Marder, D.; Tzanani, N.; Prihed, H.; Gura, S. Trace detection of explosives with a unique large volume injection gas chromatography-mass spectrometry (LVI-GC-MS) method. *Anal. Methods*, **2018**, *10*(23), 2712-2721.
[http://dx.doi.org/10.1039/C8AY00480C]

[110] Hable, M.; Stern, C.; Asowata, C.; Williams, K. The determination of nitroaromatics and nitramines in ground and drinking water by wide-bore capillary gas chromatography. *J. Chromatogr. Sci.*, **1991**, *29*(4), 131-135.
[http://dx.doi.org/10.1093/chromsci/29.4.131] [PMID: 1874908]

[111] Huri, M.A.M.; Ahmad, U.K.; Ibrahim, R.; Omar, M. A review of explosive residue detection from forensic chemistry perspective. *Malays. J. Anal. Sci.*, **2017**, *21*(2), 267-282.
[http://dx.doi.org/10.17576/mjas-2017-2102-01]

[112] Stefanuto, P.H.; Perrault, K.A.; Focant, J.F.; Forbes, S.L. Fast Chromatographic Method for Explosive Profiling. *Separations*, **2015**, *2*, 213-224.

[113] Anthony, I.G.M.; Brantley, M.R.; Gaw, C.A.; Floyd, A.R.; Solouki, T. Vacuum Ultraviolet Spectroscopy and Mass Spectrometry: A Tandem Detection Approach for Improved Identification of Gas Chromatography-Eluting Compounds. *Anal. Chem.*, **2018**, *90*(7), 4878-4885.
[http://dx.doi.org/10.1021/acs.analchem.8b00531] [PMID: 29505232]

[114] MacCrehan, W.; Moore, S.; Schantz, M. Reproducible vapor-time profiles using solid-phase

microextraction with an externally sampled internal standard. *J. Chromatogr. A,* **2012**, *1244*, 28-36.
[http://dx.doi.org/10.1016/j.chroma.2012.04.068] [PMID: 22633864]

[115] Timmer, B.; Olthuis, W.; Van Den Berg, A. Ammonia sensors and their applications—a review. *Sens. Actuators B Chem.,* **2005**, *107*(2), 666-677.
[http://dx.doi.org/10.1016/j.snb.2004.11.054]

[116] Song, L.; Bartmess, J.E. Liquid chromatography/negative ion atmospheric pressure photoionization mass spectrometry: a highly sensitive method for the analysis of organic explosives. *Rapid Commun. Mass Spectrom.,* **2009**, *23*(1), 77-84.
[http://dx.doi.org/10.1002/rcm.3857] [PMID: 19051224]

[117] Ochsenbein, U.; Zeh, M.; Berset, J.D. Comparing solid phase extraction and direct injection for the analysis of ultra-trace levels of relevant explosives in lake water and tributaries using liquid chromatography-electrospray tandem mass spectrometry. *Chemosphere,* **2008**, *72*(6), 974-980.
[http://dx.doi.org/10.1016/j.chemosphere.2008.03.004] [PMID: 18472128]

[118] Schramm, S.; Léonço, D.; Hubert, C.; Tabet, J.C.; Bridoux, M. Development and validation of an isotope dilution ultra-high performance liquid chromatography tandem mass spectrometry method for the reliable quantification of 1,3,5-Triamino-2,4,6-trinitrobenzene (TATB) and 14 other explosives and their degradation products in environmental water samples. *Talanta,* **2015**, *143*, 271-278.
[http://dx.doi.org/10.1016/j.talanta.2015.04.063] [PMID: 26078159]

[119] National Turk, Flash news, Breaking News/Suicide bomber wrecks havoc in Istanbul Taksim square, Thursday, January 9 2020. https://www.nationalturk.com/en/suicide-bomber-wrecks-havo--in-istanbul-taksim-square-356756/ January 9th, 2020. [Accessed on September, 2019].

[120] Hart-Smith, G.; Blanksby, S.J. Mass analysis.*Mass Spectrometry in Polymer Chemistry*; Barner-Kowollik, C.; Gruendling, T.; Falkenhagen, J.; Weidner, S., Eds.; Wiley-VCH Verlag & Co: Weinheim, Germany, **2012**, pp. 5-32.
[http://dx.doi.org/10.1002/9783527641826.ch1]

[121] Holčapek, M.; Jirásko, R.; Lísa, M. Recent developments in liquid chromatography-mass spectrometry and related techniques. *J. Chromatogr. A,* **2012**, *1259*, 3-15.
[http://dx.doi.org/10.1016/j.chroma.2012.08.072] [PMID: 22959775]

[122] Anilanmert, B. Liquid Chromatography– Mass Spectrometry in the Analysis of Designer Drugs.*Chromatographic Techniques in the Forensic Analysis of Designer Drugs*; Taylor & Francis, **2018**.
[http://dx.doi.org/10.1201/9781315313177-5]

[123] Ehlert, S.; Hölzer, J.; Rittgen, J.; Pütz, M.; Schulte-Ladbeck, R.; Zimmermann, R. Rapid on-site detection of explosives on surfaces by ambient pressure laser desorption and direct inlet single photon ionization or chemical ionization mass spectrometry. *Anal. Bioanal. Chem.,* **2013**, *405*(22), 6979-6993.
[http://dx.doi.org/10.1007/s00216-013-6839-8] [PMID: 23455645]

[124] Tourné, M. Developments in explosives characterization and detection. *J. Forensics Res.,* **2013**, *12*(002)

[125] Syage, J; Hanold, KA Combating Terrorism with Mass Spectrometry—Screening People for Explosives. *Spectroscopy (Springf.),* **2009**.

[126] Caygill, J.S.; Davis, F.; Higson, S.P. Current trends in explosive detection techniques. *Talanta,* **2012**, *88*, 14-29.
[http://dx.doi.org/10.1016/j.talanta.2011.11.043] [PMID: 22265465]

[127] Hagan, N.; Goldberg, I.; Graichen, A.; St Jean, A.; Wu, C.; Lawrence, D.; Demirev, P. Ion Mobility Spectrometry - High Resolution LTQ-Orbitrap Mass Spectrometry for Analysis of Homemade Explosives. *J. Am. Soc. Mass Spectrom.,* **2017**, *28*(8), 1531-1539.
[http://dx.doi.org/10.1007/s13361-017-1666-3] [PMID: 28409445]

[128] Correa, D.N.; Santos, J.M.; Eberlin, L.S.; Eberlin, M.N.; Teunissen, S.F. Forensic chemistry and ambient mass spectrometry: a perfect couple destined for a happy marriage? *Anal. Chem.,* **2016**, *88*(5), 2515-2526.
[http://dx.doi.org/10.1021/acs.analchem.5b02397] [PMID: 26768158]

[129] Forbes, T.P.; Sisco, E.; Staymates, M.; Gillen, G. DART-MS analysis of inorganic explosives using high temperature thermal desorption. *Anal. Methods,* **2017**, *9*(34), 4988-4996.
[http://dx.doi.org/10.1039/C7AY00867H] [PMID: 29651308]

[130] Tsai, C.W.; Tipple, C.A.; Yost, R.A. Application of paper spray ionization for explosives analysis. *Rapid Commun. Mass Spectrom.,* **2017**, *31*(19), 1565-1572.
[http://dx.doi.org/10.1002/rcm.7932] [PMID: 28681982]

[131] Hu, B.; So, P.K.; Chen, H. Electrospray ionization using wooden tips. *Anal. Chem.,* **2011**, *83*, 8201-8207.
[http://dx.doi.org/10.1021/ac2020273] [PMID: 21916420]

[132] Pirro, V.; Jarmusch, A.K.; Vincenti, M.; Cooks, R.G. Direct drug analysis from oral fluid using medical swab touch spray mass spectrometry. *Anal. Chim. Acta,* **2015**, *861*, 47-54.
[http://dx.doi.org/10.1016/j.aca.2015.01.008] [PMID: 25702273]

[133] Liu, J.; Wang, H.; Cooks, R.G.; Ouyang, Z. Leaf spray: direct chemical analysis of plant material and living plants by mass spectrometry. *Anal. Chem.,* **2011**, *83*(20), 7608-7613.
[http://dx.doi.org/10.1021/ac2020273] [PMID: 21916420]

[134] De Hoffmann, E. *Mass spectrometry Kirk-Othmer Encyclopedia of Chemical Technology*; John Wiley & Sons, Inc.: New York, **2000**.

[135] Ong, T.H.; Mendum, T.; Geurtsen, G.P.; Kelley, J.A.; Ostrinskaya, A.; Kunz, R.R. *Mass Spectrometry Vapor Analysis for Improving Explosives Detection Canine Proficiency*; MIT Lincoln Laboratory: Lexington, United States, **2017**.

[136] Forbes, T.P.; Sisco, E.; Staymates, M. Detection of Nonvolatile Inorganic Oxidizer-Based Explosives from Wipe Collections by Infrared Thermal Desorption-Direct Analysis in Real Time Mass Spectrometry. *Anal. Chem.,* **2018**, *90*(11), 6419-6425.
[http://dx.doi.org/10.1021/acs.analchem.8b01037] [PMID: 29701987]

[137] Bridoux, M.C.; Schwarzenberg, A.; Schramm, S.; Cole, R.B. Combined use of direct analysis in real-time/Orbitrap mass spectrometry and micro-Raman spectroscopy for the comprehensive characterization of real explosive samples. *Anal. Bioanal. Chem.,* **2016**, *408*(21), 5677-5687.
[http://dx.doi.org/10.1007/s00216-016-9691-9] [PMID: 27318472]

[138] Forbes, T.P.; Sisco, E. Mass spectrometry detection and imaging of inorganic and organic explosive device signatures using desorption electro-flow focusing ionization. *Anal. Chem.,* **2014**, *86*(15), 7788-7797.
[http://dx.doi.org/10.1021/ac501718j] [PMID: 24968206]

[139] Choi, Y.; Remmler, D.; Ries, M. Forensic analysis of explosives, Humboldt Universitat zu Berlin. https://ethz.ch/content/dam/ethz/special-interest/chab/organi-   -chemistry/zenobi-gro-   p-dam/documents/Education/LecturesExercises/Analytical%20Strategy%202014/HS2015/031115-Presentation_Explosives.pdf**2019**.

[140] Fradella, H.F.; O' Neill, L.; Fogarty, A. The impact of Daubert on forensic science. 31 Pepp. L. Rev. 2. **2004**.https://digitalcommons.pepperdine.edu/cgi/viewcontent.cgi?article=1273&context=plr

[141] Separation   Conditions2   Flow   Injection   Analysis. https://www.shimadzu.com/an/lcms/support/lib/lctalk/61/61lab.html

[142] Michel, D.; Gaunt, M.C.; Arnason, T.; El-Aneed, A. Development and validation of fast and simple flow injection analysis-tandem mass spectrometry (FIA-MS/MS) for the determination of metformin in dog serum. *J. Pharm. Biomed. Anal.,* **2015**, *107*, 229-235.
[http://dx.doi.org/10.1016/j.jpba.2014.12.012] [PMID: 25618829]

[143]   National Research Council. *Existing and potential standoff explosives detection techniques.,* **2004**,

[144]   TSA   Testing   Bomb-Detection   Scanners   At   Penn   Station.
        2018.http://newyork.cbslocal.com/2018/02/27/penn-station-tsa-explosives-scanners-testing/

[145]   Torres, N. *Applications of a portable capillary electrophoresis instrument in environmental science,*
        2015.

[146]   da Silva, L.C.; Pereira, I.; de Carvalho, T.C.; Allochio Filho, J.F.; Romão, W.; Vaz, B.G. Paper spray
        ionization and portable mass spectrometers: a review. *Anal. Methods,* **2019**, *2019*(11), 999-1013.
        [http://dx.doi.org/10.1039/C8AY02270D]

[147]   Christopher, C. Mulligan and Kyle E. Vircks, Advances in field-portable mass spectrometers for on-
        site   analytics   October   2012,   The   American   Oil   Chemists'   Society,Illinois.
        2012.https://www.aocs.org/stay-informed/inform-magazine/featured-articles/advan-

[148]   Malcolm, A.; Wright, S.; Syms, R.R.; Moseley, R.W.; O'Prey, S.; Dash, N.; Pegus, A.; Crichton, E.;
        Hong, G.; Holmes, A.S.; Finlay, A.; Edwards, P.; Hamilton, S.E.; Welch, C.J. A miniature mass
        spectrometer for liquid chromatography applications. *Rapid Commun. Mass Spectrom.,* **2011**, *25*(21),
        3281-3288.
        [http://dx.doi.org/10.1002/rcm.5230] [PMID: 22006391]

[149]   Li, L.; Chen, T.C.; Ren, Y.; Hendricks, P.I.; Cooks, R.G.; Ouyang, Z. Mini 12, miniature mass
        spectrometer for clinical and other applications--introduction and characterization. *Anal. Chem.,* **2014**,
        86(6), 2909-2916.
        [http://dx.doi.org/10.1021/ac403766c] [PMID: 24521423]

[150]   Guo, Q.; Gao, L.; Zhai, Y.; Xu, W. Recent developments of miniature ion trap mass spectrometers.
        Chin. Chem. Lett., **2018**, *29*(11), 1578-1584.--> [J]. [http://dx.doi.org/10.1016/j.cclet.2017.12.009].
        [http://dx.doi.org/10.1016/j.cclet.2017.12.009]

[151]   Snyder, D.T.; Pulliam, C.J.; Ouyang, Z.; Cooks, R.G. Miniature and fieldable mass spectrometers:
        recent advances. *Anal. Chem.,* **2016**, *88*(1), 2-29.
        [http://dx.doi.org/10.1021/acs.analchem.5b03070] [PMID: 26422665]

[152]   Xu, W.; Manicke, N.E.; Cooks, G.R.; Ouyang, Z. Miniaturization of mass spectrometry analysis
        systems, JALA. *JALA Charlottesv Va,* **2010**, *15*(6), 433-439.
        [http://dx.doi.org/10.1016/j.jala.2010.06.004] [PMID: 21278840]

[153]   Fedick, P.W.; Fatigante, W.L.; Lawton, Z.E.; O'Leary, A.E.; Hall, S.; Bain, R.M. A low-cost,
        simplified platform of interchangeable, ambient ionization sources for rapid, forensic evidence
        screening on portable mass spectrometric instrumentation. *Instruments,* **2018**, *2*(2), 5.
        [http://dx.doi.org/10.3390/instruments2020005]

[154]   Rahul, Narayanan; Depanjan, Sarkar; Graham, Cooks R; Thalappil, Pradeep Molecular Ionization
        from Carbon Nanotube Paper *Angew Chem ,* **2014**, *53*(23), 40-5936. Int. Ed. Engl. 2

[155]   Jovanic, P.B. Analysis of Explosive Residues in Soil 1st International Conference on Hazardous Waste
        Management. Chania, Greece. **2008**.

[156]   Chinchole, R.; Hatre, P.M.; Desai, U.; Chavan, R. Recent applications of hyphenated liquid
        chromatography techniques in forensic toxicology: a review. *Int. J. Pharm. Sci. Rev. Res.,* **2012**, *14*,
        57.

[157]   Sarkar, D.; Som, A.; Pradeep, T. Catalytic Paper Spray Ionization Mass Spectrometry with Metal
        Nanotubes and the Detection of 2,4,6-Trinitrotoluene. *Anal. Chem.,* **2017**, *89*(21), 11378-11382.
        [http://dx.doi.org/10.1021/acs.analchem.7b02288] [PMID: 28985051]

[158]   Cheng, S.C.; Jhang, S.S.; Huang, M.Z.; Shiea, J. Simultaneous detection of polar and nonpolar
        compounds by ambient mass spectrometry with a dual electrospray and atmospheric pressure chemical
        ionization source. *Anal. Chem.,* **2015**, *87*(3), 1743-1748.
        [http://dx.doi.org/10.1021/ac503625m] [PMID: 25562530]

[159]   Usmanov, D.T.; Chen, L.C.; Yu, Z.; Yamabe, S.; Sakaki, S.; Hiraoka, K. Atmospheric pressure chemical ionization of explosives using alternating current corona discharge ion source. *J. Mass Spectrom.,* **2015**, *50*(4), 651-661.
[http://dx.doi.org/10.1002/jms.3552] [PMID: 26149109]

[160]   Fletcher, C.; Sleeman, R.; Luke, J.; Luke, P.; Bradley, J.W. Explosive detection using a novel dielectric barrier discharge ionisation source for mass spectrometry. *J. Mass Spectrom.,* **2018**, *53*(3), 214-222.
[http://dx.doi.org/10.1002/jms.4051] [PMID: 29212136]

[161]   Poltash, M.L.; McCabe, J.W.; Patrick, J.W.; Laganowsky, A.; Russell, D.H. Development and Evaluation of a Reverse-Entry Ion Source Orbitrap Mass Spectrometer. *J. Am. Soc. Mass Spectrom.,* **2019**, *30*(1), 192-198.
[http://dx.doi.org/10.1007/s13361-018-1976-0] [PMID: 29796735]

[162]   Eaton, P.E.; Gilardi, R.L.; Zhang, M.X. Polynitrocubanes: Advanced High-Density, High-Energy Materials. *Adv. Mater.,* **2000**, *12*(15), 1143-1148.
[http://dx.doi.org/10.1002/1521-4095(200008)12:15<1143::AID-ADMA1143>3.0.CO;2-5]

[163]   Lu, M.; Zhao, G.J. Theoretical studies on high energetic density polynitroimidazopyridines. *Braz Chem Soc,* **2013**, *24*(6), 1018-1026.
[http://dx.doi.org/10.5935/0103-5053.20130131]

[164]   Fried, L.E.; Manaa, M.R.; Lewis, J.P. *Modeling the Reactions of Energetic Materials in the Condensed Phase* UCRL-BOOK-201405; Lawrence Livermore National Lab.(LLNL), Livermore, CA: **2003**.

[165]   Chen, S.L.; Yang, Z.R.; Wang, B.J. Molecular perovskite high-energetic materials. *Sci China Mater,* **2018**, *61*(8), 1123-1128.
[http://dx.doi.org/10.1007/s40843-017-9219-9]

[166]   Zhao, G.; Lu, M. A Theoretical investigation of a potential high energy density compound 3, 6, 7, 8-tetranitro-3, 6, 7, 8-tetraaza-tricyclo [3.1. 1.1 (2, 4)] octane. *Quim. Nova,* **2013**, *36*(4), 513-518.
[http://dx.doi.org/10.1590/S0100-40422013000400005]

[167]   Richard, R.M.; Ball, D.W. B3LYP calculations on the thermodynamic properties of a series of nitroxycubanes having the formula C8H8-x(NO3)(x) x=1-8. *J. Hazard. Mater.,* **2009**, *164*(2-3), 1595-1600.
[http://dx.doi.org/10.1016/j.jhazmat.2008.09.078] [PMID: 18977597]

[168]   Nagana Gowda, G.A.; Djukovic, D.; Bettcher, L.F.; Gu, H.; Raftery, D. NMR-guided mass spectrometry for absolute quantitation of human blood metabolites. *Anal. Chem.,* **2018**, *90*(3), 2001-2009.
[http://dx.doi.org/10.1021/acs.analchem.7b04089] [PMID: 29293320]

[169]   Godejohann, M.; Heintz, L.; Daolio, C.; Berset, J.D.; Muff, D. Comprehensive non-targeted analysis of contaminated groundwater of a former ammunition destruction site using 1H-NMR and HPLC-SPE-NMR/TOF-MS. *Environ. Sci. Technol.,* **2009**, *43*(18), 7055-7061.
[http://dx.doi.org/10.1021/es901068d] [PMID: 19806741]

SUBJECT INDEX

www.ingramcontent.com/pod-product-compliance
Lightning Source LLC
Chambersburg PA
CBHW050838220326
41598CB00006B/396